An Introduction to
Behavioural Ecology

An Introduction to Behavioural Ecology

J.R. Krebs Lecturer in
Zoology at the Edward Grey
Institute of Field Ornithology,
Department of Zoology, Oxford
and Fellow of Wolfson College

N.B. Davies Demonstrator in
Zoology at the
University of Cambridge
and Fellow of
Pembroke College

Drawings by Jan Parr

Blackwell Scientific Publications

OXFORD LONDON

EDINBURGH BOSTON MELBOURNE

© 1981 by Blackwell Scientific Publications
Editorial offices:
Osney Mead, Oxford, OX2 0EL
8 John Street, London, WC1N 2ES
9 Forrest Road, Edinburgh, EH1 2QH
52 Beacon Street, Boston, Massachusetts 02108, USA
99 Barry Street, Carlton, Victoria 3053, Australia

First published 1981
Reprinted 1982 (twice), 1983

Distributed in the U.S.A. by
Sinauer Associates, Inc., Publishers
Sunderland, MA 01375

Library of Congress Cataloging in
Publication Data

Krebs, John R
    An introduction to behavioural ecology.

    Bibliography: p.
    Includes index.
    1. Behaviour evolution. 2. Animals, Habits and
behaviour of. 3. Human Behaviour. 4. Animal ecology.
5. Human ecology. I. Davies, Nicholas B., joint
author. II. Title.
QH371.K73   591.51   80-25843
ISBN 0-87893-431-6
ISBN 0-87893-432-4 (pbk.)

Phototypeset by Tradespools Ltd
Frome, Somerset and
Printed and bound in Great Britain

Contents

7 Parental Care and Mating Systems, 135

8 Alternative Strategies, 153

Acknowledgements

This book is based on lectures given by us at Cambridge and Oxford Universities and we thank our students for providing a stimulating and critical audience.

We thank especially Tim Birkhead for reading the first draft of the whole manuscript and the following for comments on particular chapters: Anthony Arak, Patrick Bateson, Jane Brockmann, Tim Clutton-Brock, William Foster, Peter de Groot, Paul Harvey, Geoff Parker.

For sending manuscripts and allowing us to quote unpublished work we thank: Jeffrey Baylis, Lew Oring, Richard Wrangham, Robert Hinde, Dan Rubenstein, Peter de Groot, Uli Reyer, Bob Metcalf, Ron Ydenberg, Ric Charnov, Haven Wiley, Clive Catchpole and Malte Andersson.

Finally we thank Robert Campbell for his encouragement and enthusiasm during the preparation of this book.

Introduction

This brief introduction describes the organisation and contents of our book. The book is about the survival value of behaviour. We call this subject 'behavioural ecology' because the way in which behaviour contributes to survival and reproduction depends on ecology. If, for example, we want to answer the question 'How does living in a group contribute to an individual's survival?', we have to start thinking in terms of the animal's ecology; the kind of food it eats, its enemies, its nesting requirements and so on. These ecological pressures will determine whether grouping is favoured or penalised by selection. Behavioural ecology is not only concerned with the animal's struggle to survive by exploiting resources and avoiding predators, but also with how behaviour contributes to reproductive success. Much of the book is therefore about competition between individuals for the chance to reproduce and pass on their genes to future generations.

The book emphasises the theoretical background to each subject discussed, but we prefer to illustrate the theory with examples after a very brief general introduction, rather than developing long, abstract, theoretical arguments. Although none of the ideas we have used are difficult to understand we have placed some of the more complicated arguments and details in boxes which can be ignored if the reader is in a hurry.

Chapter 1 is a general introduction to the book, in which we distinguish between different kinds of questions that one can ask about behaviour. In particular we emphasise the difference between questions about survival value or function and those concerned with causal mechanisms. In this chapter we also introduce our view of the modern Darwinian theory of evolution as used throughout the following chapters. In this discussion we include a description of the theory of kin selection, which is important later in the book for understanding the survival value of altruistic behaviour in animals.

The next two chapters (2 and 3) outline two very different and complementary ways of studying survival value that will be used later on in the book. The method in chapter 2 is comparison between species. The rationale here is that differences between species in behaviour can be correlated with differences in their ecology. From these correlations inferences can be drawn about the adaptive significance of behavioural traits. We illustrate this approach with reference to social groups in primates. In chapter 3 we focus on the individual. Animals are viewed as making 'decisions' between alternative courses of action and the decisions can be analysed in terms of their costs and benefits. A powerful tool in this approach is optimality theory, which allows us to test hypotheses about the importance of various costs and benefits by predicting their effects on the animal's decision rules. Most of our examples refer to foraging behaviour, where this approach has been most successfully applied. Chapter 4 continues with the question of how animals exploit food resources and considers why some species feed in groups while others defend exclusive territories. We use both comparative studies and optimality models to tackle the question.

The theme of defence of resources is developed further in chapter 5 where we examine how animals compete for resources such as food, territories in good habitats, or mates. We introduce the idea of game theory as a tool for analysing how individuals behave in contests for resources.

Chapters 6, 7 and 8 are concerned with sexual reproduction. A consideration of the basic differences between males and females leads to the idea that members of one sex (usually male) may compete for access to members of the other (chapter 6). This is the theory of sexual selection. The differences between male and female also suggest that the interests of the two sexes during reproduction often differ (the theory of sexual conflict). Chapter 7 discusses how these battles within and between the sexes are influenced by ecology. Here we rely heavily on the comparative approach, correlating differences between species in sexual strategies with differences in ecology. From differences between species we turn to differences between individuals (chapter 8). We introduce the idea that different individuals within a species sometimes adopt different sexual strategies. These differences may be related to age or size, or they may simply be equally profitable, alternative, ways of achieving the same end.

In chapters 9 and 10 we take up the problem, introduced in chapter 1, of the evolution of altruistic behaviour. We illustrate the theoretical arguments with reference to 'helpers', individuals that help others to rear young instead of producing their own. Chapter 9 deals with birds, mammals, and fish and chapter 10 is devoted entirely to the social insects, where helping reaches its most sophisticated level of development. Many of the earlier chapters refer to communication as a behavioural mechanism of competition for resources and social interaction. In chapter 11 we tie together these threads in a general discussion of animal signals. We follow the pattern set in earlier chapters by considering both ecological constraints and intraspecific selection pressures.

Chapter 12 extends our discussion from intra- to interspecific interactions. We use the example of plant–pollinator and predator–prey relationships to illustrate some general principles of coevolution. At the same time we recall some of the ideas raised in earlier chapters, for example by using optimality models to predict how bees exploit flowers for their nectar and how flowers exploit bees for pollination services.

In the final chapter we do two things. First, we critically reassess the views developed in chapter 1, that the survival value of behaviour can be understood within a neo-Darwinian framework using methods such as optimality models and game theory. Second, we point out some of the ways in which studies of survival value can shed light on other kinds of questions about behaviour, such as those about ontogeny and motivation. Finally a word about the style of presentation. We generally use convenient and informal shorthand rather than traditional formal scientific style. A phrase such as 'offspring are selected to demand more food than the parent wants to give' is short for 'During the course of evolution selection acting on genetic differences in the begging behaviour of offspring will have favoured an increase in the intensity of begging. This increase will have been favoured to the extent where the level of begging by any individual offspring exceeds the optimum level for the parent.'

Some readers may wonder whether our informal shorthand, together with catchy descriptive labels for various behaviour patterns such as 'rape', 'manipulation' and 'sneaker', are a sign of sloppy thinking. There is no doubt that loose terminology almost invariably indicates imprecise thinking and half-formulated ideas. But it is equally easy to conceal woolly arguments behind an obfuscating screen of scientific jargon. We have used a simple direct style in order to make our arguments clear and not because behavioural ecology is a woolly subject. This point is nowhere better illustrated than by George Orwell in his brilliant essay 'Politics and the English Language' (1946). He translates the following well-known verse from *Ecclesiastes*, into modern English. 'I returned and saw under the sun, that the race is not always to the swift, nor the battle to the strong, neither yet bread to the wise, nor yet riches to men of understanding, nor yet favour to men of skill; but time and chance happeneth to them all.'

And now the translation.

'Objective consideration of contemporary phenomena compels the conclusion that success or failure in competitive activities exhibits no tendency to be commensurate with innate capacity, but that a considerable element of the unpredictable must invariably be taken into account.'

This translation is not only tired and ugly, lacking the fresh, vivid imagery of the biblical passage, but it replaces precise illustrations with woolly generalisation. While we cannot hope to emulate the clarity and brilliance of the writer of *Ecclesiastes*, or indeed of George Orwell, we hope we have avoided the worst excesses of the Orwellian parody and presented our ideas in simple but precise language.

Chapter 1. Natural Selection and Behaviour

Questions about behaviour

In this book we will explore the relationships between animal behaviour, ecology and evolution. We shall describe how animals behave under particular ecological conditions and then ask why has this behaviour evolved? For example, we shall attempt to understand why some animals are solitary while others go around in groups and why most individuals court before they copulate. Why do some birds have songs consisting of pure whistles while others produce buzzes and trills? Why do foraging bumblebees usually start at the bottom flowers on a plant and move vertically upwards? We shall also ask some precise, quantitative questions such as why do sunbirds defend territories containing 1600 flowers and why does the male dungfly copulate for on average 41 minutes?

Niko Tinbergen, one of the founders of ethology, emphasised that there are several different ways of answering the question 'Why?' in biology. These have come to be known as Tinbergen's four questions (Tinbergen 1963). For example, if we asked why starlings, *Sturnus vulgaris*, sing in the spring, we could answer as follows.

1 In terms of *survival value or function*. Starlings sing to attract mates for breeding.

2 In terms of *causation*; because increasing daylength triggers off changes in hormone levels in the body, or because of the way air flows through the syrinx and sets up membrane vibrations. These are answers about the internal and external factors which cause starlings to sing.

3 In terms of *development*. Starlings sing because they have learned the songs from their parents and neighbours.

4 In terms of *evolutionary history*. This answer would be about how song had evolved in starlings from their avian ancestors. The most primitive living birds make very simple sounds, so it is reasonable to assume that the complex songs of starlings and other song birds have evolved from simpler ancestral calls.

It is important to distinguish these various kinds of answer or otherwise time can be wasted in sterile debate. If someone said that swallows migrate south in the autumn because they are searching for richer food supplies while someone else said they migrated because of decreasing daylength, it would be pointless to argue about who was correct. Both answers may be right, the first is in terms of survival value or function and the second is in terms of causation. Factors influencing survival value are sometimes called 'ultimate' while causal factors are referred to as 'proximate'. It is these two answers that are the most frequently muddled up and so to make the distinction clear we will discuss an example in detail. We will also use this example to illustrate some other principles later in the chapter.

REPRODUCTIVE BEHAVIOUR IN LIONS

In the Serengeti National Park, Tanzania, lions (*Panthera leo*) live in prides consisting of between 3 and 12 adult females, from 1 to 6 adult males and several cubs. The group defends a territory in which it hunts for prey, especially gazelle and zebra. Within a pride all the females are related; they are sisters, mothers and daughters, cousins, and so on. All were born and reared in the pride and all stay there to breed. Females reproduce from the age of 4 to 18 years and so enjoy a long reproductive life.

For the males, things are very different. When they are 3 years old, young related males (sometimes brothers) leave their natal pride. After a couple of years as nomads they attempt to take over another pride from old and weak males. After a successful takeover they stay in the pride for 2 to 3 years before they, in turn, are driven out by new males. A male's reproductive life is therefore short.

The lion pride thus consists of a permanent group of closely-related females and a smaller group of separately inter-related males present for a shorter time. We will consider three interesting observations about reproductive behaviour in a pride (Bertram 1975).

1 Lions may breed throughout the year but although different prides may breed at different times, within a pride all the females tend to come into oestrus at about the same time. The mechanism, or causal explanation, is probably the influence of an individual's pheromones on the oestrus cycles of other females in the pride. A similar phenomenon occurs in schools, where girls living in the same dormitory may also synchronise their menstrual cycles perhaps due to the effect of pheromones (McClintock 1971).

The function of oestrus synchrony in lionesses is that different litters in the pride are born at the same time and cubs born synchronously survive better. This is because there is communal suckling and with all the females lactating together a cub may suckle from another female if its mother is out hunting (Fig. 1.1). In addition, with synchronous births there is a greater chance that a young male will have a companion when it reaches the age at which it leaves the pride. With a companion a male is more likely to achieve a successful takeover of another pride (Bygott, Bertram & Hanby 1980).

2 A lioness comes into heat every month or so when she is not pregnant. She is on heat for two to four days during which time she copulates once every fifteen minutes throughout the day and night. Despite this phenomenal rate of copulation the birth rate

Fig. 1.1 Selfishness and altruism in a lion pride. Top: when a new male takes over a lion pride, he kills the young cubs fathered by the previous males. Bottom: a female suckles her sister's cub alongside her own.

is low. Even for those cubs that are born, only 20 per cent will survive to adulthood. It can be calculated that there are 3000 copulations for each offspring that attains the adult stage.

The causal explanation for why lion matings are so unsuccessful is not the failure of the male to ejaculate but rather the high probability of ovulation failure by the female or a high rate of abortion. But why are females designed in this apparently inefficient way?

One hypothesis is that it may be advantageous to the female to be receptive even at times when conception is unlikely, because this means that each copulation is devalued. For a male there is only a 1:3000 chance that a given copulation will produce a surviving cub and so it is not worth fighting with other males in the pride over a mating opportunity. In fact a female may mate with all the males in a pride and they often patiently wait their turn with little sign of aggression. What looks like mating inefficiency may be a 'trick' evolved by females to maintain peace among the males and thus promote stability in the pride.

3 The importance of stability for the female arises from the fact that when a new male, or group of males, takes over a pride they sometimes kill the cubs already present (Fig. 1.1). The causal explanation for this behaviour may be the unfamiliar odour of the cubs which induces the male to destroy them. A similar effect, known as the Bruce Effect, occurs in rodents where the presence of a strange male prevents the implantation of a fertilised egg or induces abortion.

The advantage of the infanticide for the male that takes over the pride is that killing the cubs fathered by a previous male brings the female into reproductive condition again much quicker and so hastens the day that he can father his own offspring. If the cubs were left intact then the female would not come into oestrus again for 25 months. By killing the cubs she becomes ready for mating after only 9 months. It will be remembered that a male's reproductive life in the pride is short, so any individual that practices infanticide when he takes over a pride will father more of his own offspring and therefore the tendency of infanticide will spread by natural selection.

The differences between the causal and functional explanations of these three aspects of reproductive behaviour in the lions are summarised in Table 1.1.

Table 1.1 Summary of causal and functional explanations for three aspects of reproductive behaviour in lions (after Bertram 1975).

Observation	Causal explanation	Functional explanation
1. Females are synchronous in oestrus.	Chemical cues?	(a) Better cub survival (b) Young males survive better when they leave the pride if in a group.
2. During oestrus the female copulates every 15 min.	Time of ovulation concealed. Low chance of fertilisation.	Reduces male competition for mating.
3. Young die after new males take over pride.	Abortion (? chemical). Male kills young.	(a) Female comes into oestrus again quicker. (b) Male removes older cubs which would compete with his young.

Natural selection

Throughout this book we will be focusing on functional questions about behaviour. Our aim is to try and understand how an animal's behaviour is adapted to the environment in which it lives. When we discuss adaptations we are referring to changes brought about during evolution by the process of natural selection. For Charles Darwin, adaptation was an obvious fact. It was obvious to him that eyes were well designed for vision, legs for running, wings for flying and so on. What he attempted to explain was how adaptation could have arisen without a creator. His theory of natural selection, published in the *Origin of Species* in 1859, can be summarised as follows.

1 Individuals within a species differ in their morphology, physiology and behaviour (*variation*).

2 Some of this variation is *heritable*; on average offspring tend to resemble their parents more than other individuals in the population.

3 Organisms have a huge capacity for increase in numbers; they produce far more offspring than give rise to breeding individuals. This capacity is not realised because the number of individuals within a population tends to remain more or less constant over time. Therefore there must be *competition* between individuals for scarce resources such as food, mates and places to live.

4 As a result of this competition, some variants will leave more offspring than others. These will inherit the characteristics of their parents and so evolutionary change will take place by *natural selection*.

5 As a consequence of natural selection organisms will come to be *adapted* to their environment. The individuals that are selected will be those best able to find food and mates, avoid predators and so on.

When Darwin formulated his idea he had no knowledge of the mechanism of heredity. The modern statement of the theory of natural selection is in terms of genes. Although selection acts on differences in survival and reproductive success between individual organisms, or phenotypes, what changes during evolution is the relative frequency of genes. G.C. Williams (1966) made this clear with the following vivid illustration.

'Phenotypes are temporary manifestations which are the result of an interaction between genotype and environment. Socrates may have been very successful in the evolutionary sense of leaving numerous offspring but his phenotype was utterly destroyed by the hemlock and has never since been duplicated. With his death, not only did his phenotype disappear but also his genotype. Meiosis and recombination destroy genotypes as surely as death. The ultimate individual fragment that is passed on from generation to generation is the gene. Genes are potentially immortal because only they can make copies of themselves fast enough to compensate for their destruction. Although Socrates' phenotype and genotype have disappeared for ever, copies of his genes may still be with us.'

We can restate Darwin's theory in modern genetic terms as follows.

1 All organisms have genes which code for protein synthesis. These proteins regulate the development of the nervous system, muscles and structure of the individual and so determine its behaviour.

2 Within a population many genes are present in two or more alternative forms, or alleles, which code for slightly different forms of the same protein. These will cause differences in development and so there will be variation within a population.

3 There will be competition between the alleles of a gene for a particular site (locus) on the chromosomes.

4 Any allele that can make more surviving copies of itself than its alternative will eventually replace the alternative form in the population. Natural selection is the differential survival of alternative alleles.

Although selection acts ultimately on genes, the genes will be selected through a complex interaction with their environment, including their genetic environment (other genes in the gene pool with which they associate) and ecological environment (climate, predators, competitors, etc.). The individual can be regarded as a temporary vehicle or survival machine by which genes survive and are replicated (Dawkins 1976). Because selection of genes is mediated through phenotypes, the most successful genes will be those which promote most effectively an individual's survival and reproductive success (and that of relatives, see later). As a result of gene selection we would therefore expect individuals to behave so as to promote gene survival.

Before we discuss how thinking about genes can help us to understand the evolution of behaviour, we should examine the evidence that gene differences can cause differences in behaviour.

Genes and behaviour

A famous example of natural selection in action is industrial melanism in insects (Kettlewell 1973). About a hundred years ago the commonest type of peppered moth (*Biston betularia*) in Britain was pale in colour but a black form occurred occasionally as a rare genetic mutant. Today the black form has become very common in industrial parts because it is better concealed from bird predators than the pale form when resting on trees blackened by pollution. We can refer to the black colour in these moths as an adaptation against predators. The increase in the black form is a result of natural selection.

Just as an animal's morphology, such as its colour, can evolve through natural selection so can its behaviour. It is clearly adaptive for a moth to prefer a background on which it will be inconspicuous. If a melanic individual preferred to settle on pale tree trunks it would quickly be seen and eaten by birds. Experiments have shown that the moth's behaviour is also adapted to its environment; melanic individuals prefer to land on dark backgrounds (Sargent 1968; Kettlewell & Conn 1977).

Genes will determine the behaviour of an animal by coding for the chemicals that are made in its body. These will influence the development of the nervous system and muscles which control the animal's behaviour. For example genes may influence a moth's preference for a particular background simply by the way they affect its response to different light intensities, perhaps through coding for certain visual pigments or neural circuitry. We can still describe these as genes for background preference, however simply they exert their effect on behaviour.

It is important to realise that when we talk about 'genes for' a particular behaviour or 'genes for' a particular colour, we do not mean that one gene alone codes for the behaviour or structure. Genes work in concert, not in isolation. Many genes together will determine the colour of a moth and its behaviour in preferring to settle on certain backgrounds. However, just as a *difference* in colour between two individuals may be due to a *difference* in one gene (e.g. melanism in peppered moths), so a difference in behaviour can also be due to a difference in one gene. A useful analogy is in the baking of a cake; a difference in one word of a recipe may mean that the taste of the whole cake is different but this does not mean that the one word is responsible for the entire cake (Dawkins 1979). Therefore, whenever

---

Box 1.1

The differences in behaviour between the hygenic and non-hygenic strains of honeybees is due to a difference in two genes. One gene (U) controls the uncapping of the cells and the other gene (R) controls removal of the larvae.

Hygenic strains have two copies of the recessive allele of each gene (uurr).

Non-hygenic strains are UURR.

When Rothenbuhler crossed the two strains all the F1 hybrids were non-hygenic (UuRr).

When these F1 hybrids were backcrossed with the pure hygenic strain (uurr) four different genotypes were produced.

F1 hybrid ♀ UuRr

Gametes	ur	UR	uR	Ur
Hygenic strain ♂ ur	uurr*	UuRr†	uuRr‡	Uurr§

*A quarter of the offspring were hygenic (uurr), both uncapping the cells and removing the larvae.
†A quarter were non-hygenic, showing neither behaviour (UuRr)
‡A quarter uncapped the cells, but then failed to remove the larvae (uuRr)
§A quarter would remove the larvae, but only if Rothenbuhler first uncapped the cells for them (Uurr).

we talk about 'genes for' certain traits this is a shorthand for gene differences bringing about differences in structure or behaviour.

A classic example of how single gene differences may cause differences in behaviour is W.C. Rothenbuhler's (1964) work on honeybees (*Apis mellifera*). Some strains of bee are hygienic; the workers remove any diseased larvae from the nest and so prevent the spread of infection. This process involves two types of behaviour, first uncapping the wax cells and second throwing out the larvae. Other strains are non-hygienic and never throw out diseased larvae. By genetic crosses Rothenbuhler showed that the difference in behaviour between the hygienic and non-hygienic strains was due to a difference in two genes (see Box 1.1). Once again, these genes could exert their effects on behaviour by simple pathways. For example, the gene for uncapping may act just by changing the taste thresholds of the bee so that it eats infected wax.

There are also other ways of showing that genes influence behaviour. For example, Aubrey Manning (1961) was able to select for two different mating speeds in the fruit fly (*Drosophila melanogaster*) by breeding from fast and slow maters. Seymour Benzer (1973) has used mutagens (radiation or chemicals) to produce genetic mutations that change behaviour. In one mutant, known as 'stuck', the male fails to disengage from the female after the normal 20 min period of copulation. Mutation of another gene produces 'coitus interruptus' males which disengage after only 10 min and fail to produce any offspring. Benzer was able to trace the cause of these mutations and show that they resulted from abnormalities in the sensory receptors, nervous system or muscles of the flies.

It is clear that genetic differences between individuals (often differences in a single gene) can result in complex differences in behaviour simply through their effects on the development of the nervous and muscle systems. No-one has studied the genetic basis for copulation time in the male dungfly but it seems reasonable to assume that this is a result of natural selection. In a sense all behaviour must be coded for by genes; reduced to its simplest form behaviour is nothing more than a series of nervous impulses and muscle contractions and the protein structure of nerve and muscle is coded for by genetic instruction. We expect copulation time, foraging behaviour and other forms of behaviour to have been selected during evolution just like other traits such as colour.

Later on in this chapter we will talk about genes for altruism. Again, it is true that no-one has yet isolated genetic mutants showing differences in altruistic behaviour but genes for altruism are no more difficult to imagine than genes for mating speed. All we mean by a 'gene for altruism' is that an individual with the gene will behave more altruistically than other individuals with alternative alleles. Like the other examples we have discussed above, the gene may exert its effect in a very simple way, for example by causing the animal to delay eating when it has captured a prey item. If this means that other individuals are then able to eat more of the food, then the gene, in effect, is a gene for altruism.

Summarising this section, gene differences can result in behavioural differences because genes code for enzymes which influence the development of the sensory, nervous and muscle systems of the animal, which in turn affect its behaviour. In some cases single gene mutations can result in complex changes in behaviour. Even though the genetic basis for most behaviour has not been studied, it is nevertheless reasonable to assume it has evolved by natural selection.

Selfish individuals or group advantage?

As we discussed above, one way of studying animal behaviour is to ask how it contributes to an individual's survival and reproductive success. We shall see in the next section that we will have to refine this question slightly, but before we do so it is worth considering whether behaviour can evolve because it is for the good of the group. It may seem obvious from our examples so far that traits evolve not for any group benefit but because they are advantageous to the individuals who possess them. It's not to the group's advantage to have a cub killed when a new male takes over a lion pride. It's not to the lioness's advantage either, but because she is smaller than the male there is probably not much she can do about it. Infanticide has evolved because it is to the advantage of the male that practices it.

A few years ago, however, many people thought that animals behaved for the good of the group, or of the species. It was common to read (and sometimes still is) explanations like, 'lions rarely fight to the death because if they did so, this would endanger survival of the species' or, 'salmon migrate thousands of miles from the open ocean into a small stream where they spawn and die, killing themselves with exhaustion to ensure survival of the species.'

Because 'group thinking' is so easy to adopt, it is worth going into a little detail to examine why it is the wrong way to frame evolutionary arguments.

The main proponent of the idea that animals behave for the good of the group was V.C. Wynne-Edwards (1962). He suggested that if a population overexploited its food resources it would go extinct and so adaptations have evolved to ensure that each group of species controls its rate of consumption. Wynne-Edwards proposed that individuals restrict their birth rate to prevent overpopulation, for example by producing fewer young, not breeding every year, delaying the onset of breeding, and so on. This is an attractive idea because it is what humans ought to do to control their own populations. However there are two reasons for thinking it is unlikely to work for animal populations.

'GROUP THINK'

Imagine a species of bird in which each pair lays 2 eggs and there is no overexploitation of the food resources. Suppose the tendency to lay 2 eggs is inherited. Now consider a mutant which lays 6 eggs. Since the population is not overexploiting its food supplies, there will be plenty of food for the young to survive and the 6 egg genotype will become more common very rapidly. Just as with the peppered moth, the gene frequencies in the population will change.

Will the 6 egg type be replaced by birds that lay 7 eggs? The answer is yes, as long as individuals laying more eggs produce more surviving young. Eventually a point will be reached where the brood is so large that the parents cannot look after it as efficiently as a smaller one. The clutch size we would expect to see in nature will be the one that results in the most surviving young because natural selection will favour individuals that do the best. A system of voluntary birth control for the good of the group will not evolve because it is unstable; there is nothing to stop individuals behaving in their own selfish interests.

Wynne-Edwards realised this and so proposed the idea of 'group selection' to explain the evolution of behaviour that was for the good of the group. He suggested that groups consisting of selfish individuals died out because they overexploited their food resources. Groups that had individuals who restricted their birth rate did not overexploit their resources and so survived. By a process of differential survival of groups, behaviour evolved that was for the good of the group.

In theory this might just work, but groups must be selected during evolution; some groups must die out faster than others. In practice, however, groups do not go extinct fast enough for group selection to be an important force in evolution. Individuals will nearly always die more often than groups and so individual selection will be more powerful. In addition, for group selection to work populations must be isolated. Otherwise there would be nothing to stop the migration of selfish individuals into a population of altruists. Once they had arrived their genotype would soon spread. In nature groups are rarely isolated sufficiently to prevent such immigration. So group selection is usually going to be a weak force and probably rarely very important (Williams 1966; Maynard Smith 1976a). We will return to this again in chapter 13.

EMPIRICAL STUDIES

Apart from these theoretical objections, there is a good field evidence that individuals do not restrict their birth rate for the good of the group but in fact reproduce as fast as they can. A good example is David Lack's long term study of the great tit (*Parus major*) in Wytham Woods, near Oxford, England (Perrins 1965; Lack 1966).

In this population the great tits nest in boxes and lay a single clutch of eggs in the spring. All the adults and young are marked individually with small numbered metal rings round their legs. The eggs of each pair are counted, the young are weighed and their survival after they leave the nest is measured by retrapping ringed birds. This intensive field study involves several people working full-time on the great tits throughout the year, and it has been going on for 34 years! Most pairs lay 8 to 9 eggs (Fig. 1.2). The limit is not set by an incubation constraint because when more eggs are added the pair can still incubate them successfully. However, the parents cannot feed larger broods so well. Chicks in larger broods get fed less often, are given smaller caterpillars and consequently weigh less when they leave the nest (Fig. 1.3a). It is not surprising that feeding the young produces a limit for the parents because they have to be out searching for food from dawn to dusk and may deliver over 1000 items per day to the nest at the peak of nestling growth.

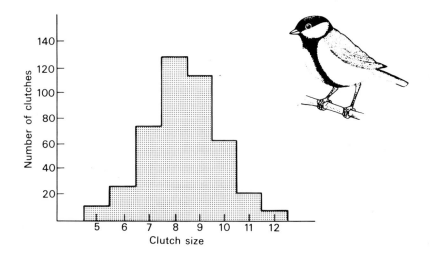

Fig. 1.2 The frequency distribution of the clutch size of great tits in Wytham Woods. Most pairs lay 8–9 eggs (Perrins 1965).

The significance of nestling weight is that heavier chicks survive better (Fig. 1.3b). An overambitious parent will leave fewer surviving young because it cannot feed its nestlings adequately. By creating broods of different sizes experimentally it can be demonstrated that there is an optimum brood size, one that maximises the number of surviving young from a selfish individual's point of view (Fig. 1.4). The commonest observed clutch size (Fig. 1.2) is close to the predicted optimum but slightly lower. This is probably because adult survival is influenced by clutch size, the strain of producing larger broods reducing the chances that the parents will survive to breed another year (Fig. 1.5; see also chapter 9). Therefore, in order to maximise total lifetime reproductive output, the parents have to be slightly conservative in their clutch size in any one season.

Our conclusion is that the clutch size is as expected from individual selection, not group selection. Individuals that lay 8 eggs predominate in the population because they give rise to most young during their

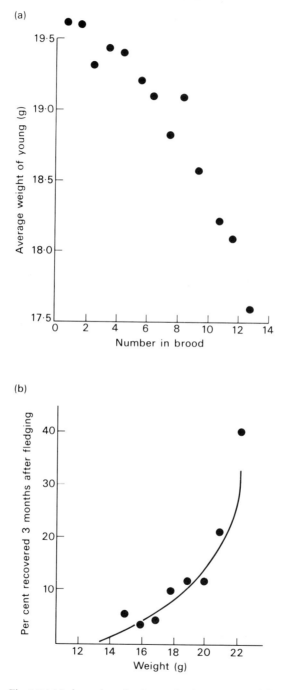

Fig. 1.3 (a) In larger broods of great tits the young weigh less at fledging because the parents cannot feed them so efficiently. (b) The weight of a nestling at fledging determines its chances of survival. Heavier chicks survive better (Perrins 1965).

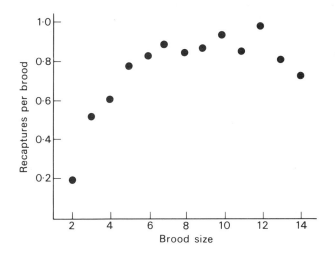

Fig. 1.4 Experimental manipulations of the number of young in a nest show that the optimal brood size for a pair of great tits is between 8 and 12 eggs. This is the brood size which maximises the number of surviving young (Perrins 1979).

lifetime, which in turn inherit the tendency to lay 8 eggs. Clutch size is maximal from the selfish individual's point of view. The exact clutch size may vary slightly from year to year and during the season, depending on food supplies, so individuals do show some variation in their birth rate. However it is always for their own selfish optima, never for the good of the group.

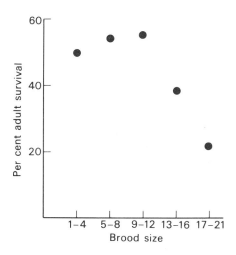

Fig. 1.5 Individual great tits rearing a large number of young within a season have a lower survival rate than those rearing fewer young (Kluyver 1971).

The evolution of altruism

We have now come down firmly in favour of individual selfishness as an expected result of natural selection and, indeed, in nature we often see animals behaving in their own selfish interests; male lions kill cubs that are not their own and thereby increase their reproductive success, great tits maximise their lifetime reproductive output, and so on.

However it is clear that individuals do not behave selfishly all the time; we also see co-operation. Animals often groom one another, individuals may join together to hunt and earlier we saw that a lioness may give suckle to another female's cubs. How can we explain the evolution of such behaviour by natural selection?

ALTRUISM BETWEEN RELATIVES

Altruism is defined as acting in the interest of others at a cost to oneself. A familiar example is, of course, parental care. We are not surprised to see a parent great tit hard at work caring for its young. This observation presents no problem for evolutionary theory because, by definition, natural selection favours individuals who maximise their gene contribution to future generations. Young great tits have copies of their parents' genes and so parental care is genotypically selfish. Just as genes for melanism in moths increase in frequency because individuals with those genes have greater reproductive success and pass copies of the genes to their offspring, so genes for altruism towards one's young will become more numerous in the gene pool because offspring have copies of those same genes. Another way of putting this is to say that the genes, by virtue of their effect on behaviour, are increasing the survival of copies of themselves in the bodies of other individuals.

We can quantify the probability that a copy of a particular gene in a parent is present in one of its offspring. In diploid species the gametes are formed by a reduction division (meiosis) in which any given gene has a 50 per cent chance of going into any one sperm or egg. There is likewise a 50 per cent chance that its allele on the homologous chromosome will go into the gamete. When an egg and a sperm fuse to produce a zygote each parent contributes exactly 50 per cent of its genes to the offspring. Therefore the probability that a parent and an offspring will share a copy of a particular gene identical by descent is 0.5. This quantity is called the *coefficient of relatedness* or r.

Now offspring are not the only relatives that share copies of the same genes identical by descent, so do other relatives. Once again we can calculate the probability that a copy of a gene in one individual will be present, by virtue of descent from a common ancestor, in a brother, sister, cousin and so on. For brothers and sisters r is 0.5, for grandchildren it is 0.25 and for cousins it is 0.125 (Box 1.2). It was

Box 1.2

Calculation of r, the coefficient of relatedness. r is the probability that a gene in one individual is an identical copy, by descent, of a gene in another individual.

General method

Draw a diagram with the individuals concerned and their common ancestors, indicating the generation links by arrows. At each generation link there is a meiosis and so a 0.5 probability that a copy of a particular gene will get passed on. For L generation links the probability is $(0.5)^L$. To calculate r, sum this value for all possible pathways between the two individuals.

$r = \Sigma(0.5)^L$.

Specific examples

These diagrams show calculations of r between two individuals represented by solid circles; other relatives are indicated by open circles. The solid lines are the generation links used in the calculations; the dotted lines are the other links in the pedigrees.

(a) Parent and offspring

$r = 1\ (0.5)^1$
$\quad = 0.5$

(b) Grandparent and grandchild

$r = 1\ (0.5)^2$
$\quad = 0.25$

(c) Full sibs (brother, sister)

$r = 2\ (0.5)^2$
$\quad = 0.5$

(Identical genes by descent can be inherited by two pathways, either mother or father)

(d) Half sibs

$r = 1\ (0.5)^2$
$\quad = 0.25$

(Identical genes by descent can only be inherited from one parent)

(e) Cousins

$r = 2\ (0.5)^4$
$\quad = 0.125$

W.D. Hamilton (1964) who realised the important implication of this for the evolution of altruism although the idea was anticipated by Fisher (1930) and Haldane (1953). Just as gene replication can occur through parental care, so it can through care for siblings, cousins or other relatives. If an individual can gain genetic representation in future generations not only by producing its own offspring but also through the reproductive success of its sisters, cousins and other relatives, then perhaps we should stop talking about selfish individuals and refer instead to selfish genes (Dawkins 1976)?

Table 1.2 shows two alternative ways of describing natural selection. Population geneticists measure gene frequencies; the unit of selection is the gene and the quantity maximised during evolution is

Table 1.2 Two alternative ways of describing natural selection (Dawkins 1978).

Unit of selection	Quantity maximised
Gene	Replication
Individual	Inclusive fitness

replication. Behavioural ecologists observe individuals and often use simply differences in reproductive success as a measure of fitness. But if our measure is to be an equivalent of the geneticists' measure of gene frequency, then we should really compute differences in a quantity that Hamilton (1964) has called '*inclusive fitness*'. This measure will reflect an individual's contribution to the gene pool and will be a sort of genetical octopus that includes the animal plus 0.5 times its number of brothers and sisters plus 0.125 times its number of cousins, and so on. We could define inclusive fitness as that property of an individual organism which will appear to be maximised when what is really being maximised is gene survival (Dawkins 1978).

Selection for behaviour that lowers an individual's own chances of survival and reproduction but raises that of a relative is known as kin selection (Maynard Smith 1964). The conditions under which an altruistic act will spread by kin selection can be specified as follows. Imagine an interaction between an altruist (or donor) and a recipient in which the costs and benefits of the interaction can be assessed in terms of offspring, or offspring equivalents. If the donor sacrifices C offspring and the recipient gains an additional B offspring, then the gene causing the donor to act will increase in frequency if

$$\frac{B}{C} > \frac{1}{r}$$

where r is the coefficient of relatedness of the donor to the recipient (Hamilton 1964). An intuitive understanding of this formula can be got as follows. As an extreme example of altruism, imagine a gene that programs an individual to die in order to save the lives of relatives. One copy of the gene will be lost from the population in the death of the altruist but the gene will still increase in frequency in the gene pool if, on average, the altruistic act saves the lives of more than two brothers or sisters ($r = 0.5$), more than four grandchildren ($r = 0.25$) or more than eight cousins ($r = 0.125$). Having made these calculations on the back of an envelope in a pub one evening, J.B.S. Haldane announced that he would be prepared to lay down his life for the sake of 2 brothers or 8 cousins!

EXAMPLES OF ALTRUISM BETWEEN RELATIVES

This illustration may seem a little far-fetched but in nature there are real examples of self sacrifice. Some insects are distasteful and warningly coloured, often with conspicuous yellow or red stripes (Fig. 1.6). A predator may come along, eat the insect, discover that it is distasteful and learn never to touch that species again. The nasty taste and warning colour are of no direct benefit to the individual that is eaten. But because its death teaches the predator a lesson, genes for the nasty taste and warning colouration will still increase in the population if the death of one or a few individuals results in decreased predation on relatives which will also carry copies of the same genes. Warningly coloured insects and other animals (e.g. tadpoles; Waldman & Adler 1979) are often aggregated in kin-groups so that the death of one individual is most likely to benefit relatives, such as siblings.

Fig. 1.6 Cinnabar moth caterpillars (*Callimorpha jacobaeae*) are distasteful and warningly coloured with orange and black bands.

Another example of extreme altruism is the evolution of sterile castes in the social insects where some females, known as workers, rarely reproduce themselves, but instead help others to raise offspring. Darwin regarded this observation as potentially damaging to his theory of natural selection. How can such altruism evolve if the altruists never reproduce? Hamilton's idea immediately suggests a possible answer to this problem because the sterile workers usually help their mothers (queens) to have offspring (see chapter 10).

Of course, not all acts of altruism are as extreme as suicide and sterility. More commonly an individual will suffer smaller costs to its own chances of survival and reproduction by the altruistic act. For example some animals give alarm calls which warn others of the approach of a predator. Paul Sherman (1977) found that when a Belding's ground squirrel, *Spermophilus beldingi*, gave an alarm call it was more likely to be attacked by a predator, but individuals nearby benefited from the early warning by retreating quickly down their burrows (Fig. 1.7). In this case the individuals that got the benefit were often relatives of the caller (sisters and sisters' offspring) and so it is possible that alarm calling has spread by kin selection.

Fig. 1.7 Belding's ground squirrels, *Spermophilus beldingi*, live in colonies. In this picture a female has seen an approaching coyote and is giving an alarm call. By calling she is increasing her risk of being attacked by the predator but the warning benefits neighbouring females who retreat down their burrows. Neighbouring females are closely related (sisters, nieces, daughters) From Sherman (1977).

For other examples of altruism we can return to the lions we discussed earlier in the chapter. In a pride, the lionesses are related to each other by an *r* of 0.15 on average, roughly the equivalent of full cousins. When a female suckles another lioness's cubs, the success of those cubs will contribute to her genetic representation in future generations. Similarly, when a lion co-operates with his brother, or half-brother, to takeover a pride, he will contribute copies of his genes to the gene pool both through his own offspring and also through those sired by his brother (Bertram 1976; Bygott, Bertram & Hanby 1979).

PROBLEMS WITH INCLUSIVE FITNESS

Hamilton's model has shown, in theory, how altruism towards kin can evolve by natural selection. Such altruism evolves by gene selfishness. However, it is important to bear in mind that the model is phrased in terms of costs and benefits to individuals and does not consider any specific genetic mechanism of selection for the altruistic allele. As we stressed earlier in the chapter, evolution brings about changes in gene frequencies, so can we use models which just consider benefits and costs to individuals to predict the evolution of altruism? Although there are differences between the conclusions of formal genetic models and inclusive fitness analysis (Maynard Smith 1981; see chapter 13 for an example) these are small compared with the accuracy with which measurements can be made in the field.

The field worker often faces enormous practical difficulties in measuring B, C and r. The coefficient of relatedness, r, is often difficult to measure in the wild because although it is usually clear who is the mother of a particular individual, it is often difficult to be certain of the father, a problem in our own species which has inspired many novels and plays and given rise to several famous court cases. We shall see later in the book (chapters 6, 7 and 8) that in many species the females are mated by several males and even in supposedly monogamous species there are frequent 'sneaky matings' that are difficult to detect. Field workers are beginning to rely less on observation and to check the degrees of relationship between individuals by isoenzyme analysis (e.g. chapter 10). Benefits and costs are no less difficult to measure. Imagine the problem of quantifying the costs and benefits in terms of survival and reproduction (e.g. offspring lost and gained) when a ground squirrel gives an alarm call to warn relatives of the approach of a predator. Nevertheless there are some examples where we can estimate these values and in chapters 9 and 10 we shall make calculations based on Hamilton's formula to predict whether individuals of some species of birds and insects will have greater genetic success by helping relatives to rear young or by breeding themselves. In many cases, however, the relevant measurements have not been made.

Despite all these problems, Hamilton's insight has had a major impact on the way we think about the evolution of behaviour. We now recognise that having one's own offspring is not the only pathway to genetic success in future generations. When we ask the question, why does this animal adopt strategy A rather than strategy B, the metric by which we should assess which strategy will be selected in evolution is the one of inclusive fitness. In helping to formulate a research program it is often useful to ask ourselves 'How does that behaviour maximise the individual's inclusive fitness?' It is also interesting to ask how animals achieve the behaviour we would predict from kin selection ideas. In many cases an individual probably uses simple rules to learn who its relatives are and so determine how it will direct its altruistic behaviour (Bateson 1978; see chapter 13).

ALTRUISM BETWEEN UNRELATED INDIVIDUALS

Not all acts of altruism are directed towards close relatives. It is useful to distinguish three ideas as to how altruism between unrelated individuals can evolve.

(a) Reciprocal altruism

When a female olive baboon, *Papio anubis*, comes into oestrus a male forms a consort relationship with her, following her around wherever she goes, awaiting the opportunity to mate. Sometimes a male who does not have a female enlists the help of another, unrelated male. This solicited male engages the consort male in a fight and while they are busy doing battle, the male who enlisted help goes off with the female (Packer 1977). On a later occasion, the roles are reversed; the male who gave help is assisted by the one who received help previously. This is an example of reciprocal altruism (Trivers 1971). Whenever the benefit of an altruistic act to the recipient is greater than the cost to the actor, then as long as the help is reciprocated at some later date, both participants will gain.

Reciprocation is common in human society and we use money to regulate its use. The problem for its evolution in animal populations is the possibility of cheating. What is to stop an individual accepting help and then refusing to repay it later? Reciprocal altruism will only evolve if there is discrimination against cheaters, if the same individuals meet repeatedly and if the benefits to the individual that is helped are large compared with the costs to the helper. In addition it can only work once there is a sufficient frequency of reciprocators in the population. Imagine, for example, a population in which there is just one altruist. It would help others but receive no help in return. Reciprocation therefore cannot account for the initial spread of altruism. However once it has spread, reciprocal altruism is evolutionarily stable (Maynard Smith 1981; see chapter 5).

(b) Mutualism and individual benefit

Many cases of apparent altruism may be best explained simply by the fact that there is an obvious individual advantage in associating with or helping others whether they are related or not. An individual may do best in a group because it can find food more efficiently and is safer from predators (chapter 4). Giving an alarm call may bring a benefit to the caller even if those who get the warning are not relatives. For example a predator may be more likely to return to places where it has been successful before. If an alarm call decreases the chances that a neighbour will be caught then the predator will be less likely to come back to the area in the future (Trivers 1971).

In some cases unrelated animals may benefit from associating together in a particular task (mutualism). Two pied wagtails, *Motacilla alba*, who join together to defend a winter feeding territory can each enjoy a greater feeding rate than they would by being alone because the benefits of the association in terms of improved territory defence outweigh the costs of having to share the food supply with another bird (Davies & Houston 1981). In other instances both kin selection and mutualism may be relevant. A single lioness is not very effective at capturing a zebra, *Equus burchelli*, but by hunting with another female she improves her capture success so much that the benefits outweigh the costs of having to share the meat once the prey is caught (Caraco & Wolf 1975). Because lionesses in a pride are related, individuals will also get kin selected benefits from the association. Similarly, a single male lion has little chance of taking over a pride. By co-operating with another male he gains an individual advantage but because the males that takeover a pride are related, kin selection will also operate. The main point is that in many of these examples of altruism it would pay individuals to co-operate even if they were unrelated. In lions there are, in fact, a few observations of unrelated males holding a pride together.

(c) Altruism and manipulation

The hypotheses to account for altruism that we have discussed so far explain its occurrence in terms of individuals maximising their inclusive fitness. The altruist benefits either by helping relatives (including offspring) or by increasing its own survival. There may, however, be examples of altruism in which the altruist makes no genetic gain, but which arise because of what might be termed manipulation. A familiar example is parental care devoted by birds such as dunnocks (*Prunella modularis*) to parasitic cuckoo nestlings. The host parents carry out the costly task of rearing the cuckoo not because of some short- or long-term genetic advantage but simply because the brood parasites have successfully exploited them. We use the term 'manipulation' because it describes the fact that cuckoos have turned the normally adaptive parental responses of the host to their own benefit. In this example the altruism is directed by members of one species (the host) to individuals of another species (the parasite), but in principle the same idea could be used as an hypothesis to account for altruism within a species. In chapter 10 we shall refer to the suggestion that sterile worker insects might have originated as a result of mothers manipulating their offspring. This example also serves to illustrate the point that the different hypotheses to account for altruism are not mutually exclusive. The sterile workers obtain some kin-selected advantage from helping to rear siblings (p 195), but if an hypothetical worker that lost its sterility and reproduced itself could do better (in terms of inclusive fitness) than one that is sterile, it might be necessary to call upon additional hypotheses, such as parental manipulation, to account for the evolution of sterility.

GENETIC PREDISPOSITIONS AND ECOLOGICAL CONSTRAINTS

Summarising our discussion of altruism, individuals are genetically predisposed to help not only their own offspring but also other relatives. Even though an individual may suffer a cost to its own

Box 1.3 Some definitions

1   Altruism. Acting in the interest of others at a personal cost in terms of chances of survival and reproduction.

2   Kin selection. Selection for a trait through its effect on close relatives. In this chapter we have distinguished between *individual selection* which refers to effects on an individual's survival and reproduction and *kin selection* which we have used to refer to effects on relatives other than offspring. Others would prefer not to distinguish between offspring and other kin and would regard care for offspring as an example of kin selection. The important point is that an individual can pass on genes through relatives other than offspring. Whichever route the genes take the consequence of selection is changes in relative gene frequencies in the gene pool. For this reason some workers prefer simply to think of *gene selection.*

3   Coefficient of relatedness (r). The probability that two individuals will share copies of a particular gene identical by descent. Under this definition r between siblings for example, is 0.5. The term *identical by descent* is important. It means that the coefficient of relatedness between two individuals refers to the probability that a shared gene is descended from the same copy of the gene in their most recent common ancestor. For siblings, only 50 per cent of their genes are identical by descent but they may share well over 90 per cent of their genes altogether. Indeed, all members of a species may share about 90 per cent of their genes. It is sometimes said that therefore we would expect individuals to be altruistic to all members of their species. This argument is fallacious for the following reason. If an aid-giver donates benefit at random to other members of the population, there is on average an equal increment in fitness to all genotypes so that no evolutionary change occurs. Gene frequencies will only change as a result of differential increments in fitness. Therefore, regardless of the overall degree of genetic similarity between members of a population, an altruistic trait will spread only if altruism is given out preferentially to close relatives (see also Dawkins 1979, for further discussion).

4   Costs and benefits of interactions between individuals. One way of characterising different kinds of interaction is by their costs and benefits, as shown in the following table.

		Net benefit for actor	
		Gain	Loss
Net benefit	Gain	Mutualism (or co-operation)	Altruism
for recipient	Loss	Selfishness (or competition)	Spite

reproductive success by acting altruistically, if the act increases the reproductive success of relatives sufficiently then the altruism can evolve by kin selection. Ecological factors such as food, predators and places to live will determine the costs and benefits associated with particular strategies. Sometimes an individual will maximise its inclusive fitness by behaving selfishly and sometimes by helping others. Definitions of some of the terms used in this section are summarised in Box 1.3.

Testing evolutionary hypotheses about adaptation

ADAPTATION AS AN ASSUMPTION

In the chapters that follow one of our main assumptions is that animals are well adapted to their environment. Adaptation can be defined as 'a difference between two phenotypic traits, or complex of traits, which increases the inclusive fitness of its carrier' (Clutton-Brock & Harvey 1979). With this definition the difference between an individual exhibiting the adaptive behaviour and one that does not could be due to a difference in learning and experience rather than a genetic difference, though presumably the ability to learn will have some genetic basis. For example some titmice, *Parus* spp., in Britain have learned to peck holes in the aluminium foil caps of milk bottles and so help themselves to the breakfast cream. This behaviour is adaptive because it helps the birds survive in winter. It is usually not possible to assess directly the effect of different traits on reproductive success and so often we recognise adaptations indirectly by the fact that the trait shows a closely designed fit to some problem presented by the animal's environment (Williams 1966).

Two points should be made about the concept of adaptation. First, not all beneficial consequences of behaviour can be said to be adaptations. Some may simply be fortuitous effects which are the result of chance rather than design by natural selection. G.C. Williams illustrates this point with the example of a fox on its way to the hen house after a heavy snowfall. On subsequent trips it will probably follow the same course and make use of the pathway it has trampled through the snow, so saving time and energy which may be crucial for survival. We would not regard fox legs as adaptations for making pathways through the snow. Their structure suggests that they are designed for running and walking. However we could regard the fox's behaviour in choosing the easiest pathway as an adaptation (see also Hinde 1975).

Second, we need to bear in mind the possibility that not all differences between traits are adaptive. Some differences could be selectively neutral or represent alternative adaptive peaks (see chapter 2). Furthermore, because the environment changes and evolution is not instantaneous, some traits may be maladaptive because they were selected in relation to previous environments. We shall refer to this possibility again later on (chapters 3 and 8).

Most of the behaviour we shall describe in this book, however, is so striking and constant within the species (e.g. infanticide in lions and alarm calling in squirrels) that it is difficult to imagine it as selectively neutral. We think that it is more constructive to regard these as adaptations, arising as a necessary consequence of natural selection.

We must stress again that the concept of adaptation is an assumption. We are not testing whether animals are adapted; rather the question we shall ask in this book is how does a particular behaviour contribute to the animal's inclusive fitness. We are interested in trying to understand the selective pressures responsible for the behaviour.

METHODS FOR TESTING HYPOTHESES

One of the dangers of asking functional questions about behaviour is that it can lead to the invention of plausible stories which are not easy to test. Many adaptive 'explanations' are more a reflection of an observer's ingenuity rather than a description of what is going on in nature. As such they have no more claim as science than Rudyard Kipling's *Just So Stories* (Gould & Lewontin 1979). How can we avoid this problem?

A rigorous scientific approach to the function of behaviour will involve four stages; observations, hypotheses, predictions and tests. The first two, observation and hypothesis, often go hand in hand. It may take many years getting to know a particular species before it is possible to ask good questions about its behaviour and ecology. Niko Tinbergen's (1953) work on the herring gull, *Larus argentatus*, was the result of over twenty years' painstaking observations of the bird's behavioural repertoire and the environment in which it lives. Having observed some aspect of the behaviour or ecology of our animal that we do not understand, the next stage is to devise alternative hypotheses and to test their predictions. Predictions can be tested in four main ways.

(a) Observations

Sometimes we will be able to test between our hypotheses simply by further observation. We may see a bird defending a territory and suggest several hypotheses about its function; for example it could be a feeding territory, a territory for attracting mates or an area defended to decrease nest predation. These hypotheses are weak ones which just make qualitative predictions and we could easily test between them by observation. If the territory was only defended in the non-breeding season, for example, we could reject the last two hypotheses. In chapters 3 and 4 we will describe how it is possible to devise stronger hypotheses which make quantitative predictions about behaviour.

(b) Comparison between individuals within a species

When David Lack considered the adaptive significance of clutch size in the great tit he looked at natural variation in the population and found that the commonest observed clutch size was close to the one which maximised the number of young fledgings from a nest. Such a comparison between individuals within a species can often give us a good idea as to the function of a particular trait. For example we may observe the frogs in a pond and find that the males with the deepest croaks attract the most mates. We may then infer that deep croaks function to attract mates. The problem with this method of investigating function is that there will often be confounding variables; the frog with the deepest croak may also have the best territory and so we would need an experimental approach to discover which was important in attracting the female. We should also recognise that natural variation within a population may often arise because different individuals have different optimal strategies. For example, some male frogs get mates by sneaky methods rather than by croaking.

(c) Experiments

Niko Tinbergen pioneered the method of elegant field experimentation to answer functional questions. For example, to test the hypothesis that spacing out of gull nests functioned to reduce predation he put out experimental plots of eggs with different spacing patterns and found that those with a clumped distribution suffered greater predation than those that were spaced out as in nature (Tinbergen et al. 1967). We will be examining the experimental approach to adaptation in detail in chapter 3.

(d) Comparison among species

Different species have evolved in relation to different ecological variables and comparison among species may help us to understand how different factors, such as food and predation, have influenced the evolution of particular traits. For example, if the European passerine species are grouped into two ecological categories, namely those that nest in holes and those that build open nests, it is found that the hole nesters lay larger clutches (Table 1.3). The same relationship occurs in ducks where open nesting species again have smaller clutches than hole nesters (Lack 1968). In holes the young are relatively safe from predators but in the open there is a premium for getting the young out of the vulnerable nest as soon as possible. The same quantity of food could be used to rear a small brood quickly or a larger brood more slowly. In open nesting species the higher risks of predation have apparently selected for a smaller clutch size and rapid nestling growth. We will consider this comparative approach to adaptation in more detail in the next chapter.

Table 1.3 A comparison of the clutch size of European passerine birds grouped into two ecological categories. Hole nesting species have larger clutches (after Lack 1968).

Nest type	Estimated per cent predation	Average clutch size	Average length of nestling period (days)
Holes	30	6.9	17
Open	70	5.1	13

Summary

Natural selection is the differential survival of alternative alleles. Selection of these alternative alleles is mediated through their effects on phenotypes. The genes that are selected will be the ones most effective in promoting an individual's ability to pass those genes on to future generations. Individuals are expected to behave in their own selfish interests rather than for the good of the group. However an animal can gain genetic representation in future generations not only by producing offspring but also by helping other relatives, who will share copies of the same genes identical by descent. Therefore altruism can evolve through kin selection. Altruism may also occur between unrelated individuals because there is reciprocation, manipulation, or simply because an individual does best by co-operating with others. Behavioural ecologists usually assume that animals are well adapted to their environment and ask how a particular behaviour contributes to an individual's inclusive fitness. Such functional questions about behaviour can be answered by observations of individuals, comparison between individuals within a species, experiments and comparison among different species.

Further reading

The books by G.C. Williams (1966) and Richard Dawkins (1976) are excellent discussions of behaviour and evolution. Williams emphasises the evolution of individual selfishness as opposed to behaviour for the good of the group. Dawkins (see also his 1978 and 1979 papers) champions the view that we should think in terms of genes rather than individuals in order to understand the evolution of behaviour.

Alexander (1974) gives a general review of the evolution of social behaviour and the volume edited by Alexander & Tinkle (1981) contains some studies of particular species and problems. The papers by Paul Sherman (1977, 1980, 1981) provide a beautiful discussion of selfishness and altruism in the Belding's ground squirrel. Alcock's (1979) book is a fine introduction to both causal and functional aspects of animal behaviour.

Chapter 2. Ecology and Adaptation: Comparison between Species

The idea of comparison lies at the heart of most hypotheses about adaptation. It is the comparative study of different species which gives us a feel for the range of strategies that animals adopt in nature. When we ask functional questions about the behaviour of particular species we are usually asking why it is different from other species. Why does species A live in groups compared with species B which is solitary? Why do males of species B mate monogamously compared with males of species A which are polygynous, and so on? It is clear from field observations that closely related species often behave very differently. One of the most likely reasons for such differences is that the species have evolved in relation to different ecological pressures such as the abundance and distribution of food and predators. A powerful method for studying adaptation is to compare groups of related species and attempt to find out exactly how differences in their behaviour reflect differences in ecology. In this chapter we will first describe two examples which pioneered the comparative approach and inspired workers to use the method with other animal groups. Then we will point out some of the methodological difficulties in formulating and testing hypotheses based on comparison. Finally we will describe some recent examples of the comparative method which have attempted to overcome these problems.

Weaver birds

The first person to attempt a systematic analysis of this kind was John Crook (1964) who studied about 90 species of weaver birds (Ploceinae). These are small finches which live throughout Africa and Asia, and although they all look rather alike there are some striking differences in their social organisation. Some are solitary, some go around in large flocks. Some build cryptic nests in large defended territories while others cluster their nests together in colonies. Some are monogamous, with male and female forming a permanent pair bond; others are polygamous, the males mating with several females and contributing little to care of the offspring. How can we explain the evolution of this great diversity in behaviour?

Crook's approach was to search for correlations between these aspects of social organisation and the species' ecology. The ecological variables he considered were the type of food, its distribution and abundance, predators and nest sites. His analysis showed that the weaver birds fell into two broad categories.

32

Chapter 2
Ecology and
Adaptation:
Comparison
between
Species

1   Species living in the forest tended to be insectivorous, solitary feeders, defend large territories and build cryptic solitary nests. They were monogamous and males and females had similar plumage.

2   Species living in the savannah tended to eat seeds, feed in flocks and nest colonially in bulky conspicuous nests. They were polygamous and there was sexual dimorphism in plumage, the males being brightly coloured and the females rather dull (Fig. 2.1).

Fig. 2.1 Top: some species of weaver birds, like this one (*Malimbus scutatus*) are insectivorous, building cryptic solitary nests in the forest and feeding alone in large territories.
Bottom: other species like this one (*Ploceus cucullatus*) feed on seeds out in the open savanna. They build conspicuous nests in colonies and feed in flocks. The males are often brightly coloured.

Why is the behaviour and morphology of the weaver birds linked to their ecology in such a striking way? Crook invoked predation and food as the main selective pressures that have influenced the evolution of social organisation. His argument was as follows.

33

*Chapter 2
Ecology and
Adaptation:
Comparison
between
Species*

1 In the forest, insect food is dispersed. Therefore it is best for the birds to feed solitarily and defend their scattered food resources as a territory. Because the food is difficult to find, both parents have to feed the young and therefore stay together as a pair throughout the breeding season. With male and female visiting the nest, both must be dull coloured to avoid attracting predators. Cryptic nests spaced out from those of neighbours decrease their vulnerability to predation.

2 In the savannah, seeds are patchy in distribution and locally superabundant. It is more efficient to find patches of seeds by being in a group because groups are able to cover a wider area in their search. Furthermore the patches contain so much food that there is little competition within the flock while the birds are feeding.

In open country the birds cannot hide their nests and so they seek safety in protected sites, such as spiny acacia trees. Nests are bulky to provide thermal insulation against the heat of the sun. Because good breeding sites are few and scattered, many birds nest together in the same tree. Within a colony, males compete for nest sites and those that defend the best sites attract several females while males in the poorer parts of the colony fail to breed. In addition, because food is abundant, the female can feed the young by herself and so the male is emancipated from parental care and can spend most of his time trying to attract more females. This has favoured brighter plumage colouration in males and the evolution of polygamy.

Supporting evidence for this interpretation comes from species with intermediate ecology. The grassland seed-eaters have patchy food supplies so group living is favoured for efficient food finding. However, in grassland the nests are vulnerable, so predation favours spacing out. The result is a compromise; these species have an intermediate social organisation, nesting in loose colonies and feeding in flocks.

These results show clearly how food and predation may be important in determining social organisation. They also reveal how several different traits such as nests, feeding behaviour, plumage colour and mating system can all be considered together as a result of the same ecological variables. Crook's work with the weaver birds inspired several people to use the comparative method to study social organisation in other groups. David Lack (1968) extended the argument to include all bird species and Peter Jarman (1974) used the same approach for the African ungulates.

African ungulates

Jarman (1974) considered 74 species of ungulates; all eat plant material but differences in the precise type of food eaten are correlated with differences in movements, mating systems and antipredator behaviour. The species were grouped into five ecological categories (Table 2.1). Just as in the weaver birds, several adaptations seem to go together.

34

Chapter 2
Ecology and
Adaptation:
Comparison
between
Species

The major correlate of diet and social organisation is body size. Small species have a higher metabolic requirement per unit weight and need to select high quality patches of food such as berries and shoots. These tend to occur in the forest and are scattered in distribution, so the small species are forced to live a solitary existence. The best way to avoid predators in the forest is to hide. Because the females are dispersed, the males must also be dispersed and the commonest mating system is for a pair to occupy a territory together.

Table 2.1 The social organisation of African ungulates in relation to their ecology (from Jarman 1974)

	Exemplary species	Body weight (kg)	Habitat	Diet	Group size	Reproductive unit	Anti-predator behaviour
Group I	Dikdik Duiker	3–60	Forest	Selective browsing; fruit, buds	1 or 2	Pair	Hide
Group II	Reedbuck Gerenuk	20–80	Brush, riverine grasssland	Selective browsing or grazing	2 to 12	Male with harem	Hide, flee.
Group III	Impala Gazelle Kob	20–250	Riverine woodland, dry grassland	Graze or browse	2 to 100	Males territorial in breeding season	Flee, hide in herd
Group IV	Wildebeest Hartebeest	90–270	Grassland	Graze	Up to 150 (thousands on migration)	Defence of females within herd	Hide in herd, flee.
Group V	Eland Buffalo	300–900	Grassland	Graze unselectively	Up to 1000	Male dominance hierarchy in herd	Mass defence against predators

At the other extreme, the largest species eat poor quality food in bulk and graze less selectively on the plains. It is not economical to defend such food supplies and these species wander in herds, following the rains and fresh grazing. In these large herds there is potential for the strongest males to monopolise several females by defence of a harem or a dominance hierarchy of mating rights. When predators come along these species cannot hide on the open plains and so either flee or rely on safety in numbers in the herd. Ungulates of intermediate size show aspects of ecology and social organisation in between these two extremes (Table 2.1).

Adaptation or story-telling?

This comparative approach to adaptation is persuasive, but there are problems (Clutton-Brock & Harvey 1979; Gould & Lewontin 1979). Many of the following problems are not unique to comparative studies and it is worth bearing them in mind throughout the book.

STORY-TELLING

The explanations for the differences in social organisation are invented to fit the facts. They are plausible, but alternative hypotheses have not been considered. It is rather like observing that flamingoes are pink and then proposing that this is so they can blend in with the setting sun. This may sound silly, but how can we be sure that the ideas put forward above are any better?

35
Chapter 2
Ecology and
Adaptation:
Comparison
between
Species

The ecological variables, such as predator pressure and patchy environment, have also been used in rather a vague way. Indeed, in the weaver birds we invoked a patchy food distribution as responsible for the evolution of flocking while in the ungulates we said that high quality patchy foods favour a solitary existence. There is the obvious danger here of explaining things too easily by handwaving, by thinking up explanations to fit the facts without any rigorous quantification of the ecological factors concerned.

CONFOUNDING VARIABLES

Consider the observation that weaver birds with a diet of seeds go about in flocks. Our explanation was that seed-eating selects for flocking because this is the best way to find a patchy food supply. However we could equally well have suggested that predation selects for flocking and, as a consequence, the birds are forced to select locally abundant food so all the flock can get enough to eat. In this case a diet of seeds is a consequence, or effect, of flocking, not a cause. Maybe predation also selects for flocking in the forest insectivores but because their diet is incompatible with flocking they have to forage singly.

This is really an historical problem equivalent to that of which came first, the chicken or the egg? It means that with the comparative approach there will often be confounding variables. For example, we observe the giraffe with its long neck feeding at the tops of trees and the buffalo with its short neck feeding on the ground. We then say that a long neck is an adaptation for feeding high up in the trees. But a long neck could equally well aid predator detection. How can we control for such confounding variables and decide which selective pressure has favoured the trait? Perhaps it is both?

A particularly important confounding variable in comparative studies is body size. Jarman controlled for this in his analysis of the ungulates by dividing the species up into categories of different body weight (Table 2.1). Most biological traits do not increase in a 1:1 relationship with body size; their relation to body size is said to be allometric (Gould 1966). For example, the brain mass of different bird species increases at about the two-thirds power of body weight. In this case, before we can examine the ecological correlates of brain size we first have to remove the effects of body size. This can be done by calculating the appropriate line of best fit when brain mass is plotted against body weight and then measuring deviations from the line to see whether the size of the trait is greater or less than expected from body weight.

36

*Chapter 2
Ecology and
Adaptation:
Comparison
between
Species*

ALTERNATIVE ADAPTIVE PEAKS OR NON-ADAPTIVE
DIFFERENCES

It is tempting when comparing between species to assume that differences are always adaptive but some differences may simply be alternative solutions to the same ecological pressures. An ecologist from Mars who visited the Earth would observe that in the United States people drive their cars on the right hand side of the road while in Britain they drive on the left. He would then perhaps make lots of measurements in an attempt to find ecological correlates to explain the adaptive significance of the difference. In fact driving on the right and driving on the left may just be equally good alternatives for preventing accidents (Dawkins 1980).

Some differences between animals may be like this. Sheep use horns for fighting and deer use antlers. Horns are derived from skin while antlers are derived from bone (Modell 1969). The differences between horns and antlers need not necessarily reflect ecological differences; it may simply be a case of evolution working with different raw materials to produce the same functional end (Fig. 2.2). The problem

Fig. 2.2 The horns of the sheep (left) and antlers of the deer (right) are both used in fighting. Horns are derived from skin and antlers from bone.

with non-adaptive explanations is that they are hypotheses of the last resort. Further scientific enquiry is stifled. Maybe there is an adaptive explanation for the difference but we just haven't discovered it yet. For example, antlers are dropped and then renewed each year whereas horns are not. Perhaps this difference is related to the extent of seasonal variation in mating competition and food supply?

These criticisms are important, but they certainly do not mean that the comparative method is a failure. On the contrary, the approach is impressive in the way it brings together such a wide diversity of behavioural and morphological traits within the same ecological framework. Crook's study of the weaver birds and Jarman's work on the antelopes have served as models for ecological work on other

(a)

(b)

(c)

(d)

Fig. 2.3 Four photographs illustrating the variety of primate social organisation. (a) Table 2.2, Grade 1: A solitary insectivorous prosimian, the lesser bushbaby (*Galago senegalensis*) (photo by Caroline Harcourt). (b) Table 2.2, Grade II. A monogamous pair of folivorous, arboreal, black gibbons (*Hylobates concolor*). The male is on the left. © Ron Tilson/BPS. (c) Table 2.2, Grade IV. Part of a troop of folivorous, arboreal, dusky langurs (*Presbytes obscurus*). © Ron Tilson/BPS. (d) Table 2.2, Grade V. A troop of savannah-dwelling olive baboons (*Papio anubis*). There are two subordinate males in the foreground and the dominant males are in the background near the females and young. Photo supplied by Anthro-Photo, Photographer Irven DeVore.

groups of species. However the most recent comparative studies have attempted to control for these various problems, and we will now discuss another example, bearing the criticisms in mind, to illustrate how changes in methodology have made comparison between species a more rigorous exercise.

Primate social organisation

Early knowledge of primate behaviour came mainly from studies in zoos. In 1932 Lord Zuckerman suggested that primates tend to be social animals because they have continuous sexual activity. It should be clear from the last chapter that this is a causal explanation and leaves the functional significance of sociality unexplained. In the 1950s, the first studies of primates in the field (e.g. Carpenter 1954) revealed that sexual activity was not in fact continuous. It also became clear that different species have very different social organisations (Fig. 2.3). Tiny tarsiers and lemurs hunt solitarily in the tree tops for insects at night. Some monkeys go around in small groups in the trees by day, feeding on leaves or fruit. Others are terrestrial and live in large troops. Among the apes, the orangutan is solitary, the gibbon lives in pairs and small family groups while the chimpanzee may live in bands of up to 50.

How can we explain the evolution of this bewildering array of social organisation? It soon became apparent that ecological factors were important. For example, DeVore (1965) noticed that, compared with other species of primates, anubis baboons live in large groups, the males are large and they have big teeth. He suggested these may all be adaptations to predator defence in a terrestrial environment. By 1966, there were sufficient field data for John Crook and Stephen Gartlan to apply the first comparative approach to a large number of primates.

Like the weaver bird and antelope work, they categorised the species into several groups based on ecology and behaviour. Table 2.2 shows that at one extreme, insectivorous primates are nocturnal, forest animals which are solitary; then there is a variety of fruit and leaf-eating species which are diurnal and live in small to large groups; finally, at the other extreme, are the vegetarian browsers of the open country which live in large groups and show intensive competition between males for females and marked sexual dimorphism.

Once again, food and predation were suggested as the main selective pressures responsible for this link between social organisation and habitat. Insects are dispersed and difficult to find, so just like the insectivorous weaver birds, these primates are solitary. In open country, predation favours grouping for safety and food is locally abundant so this allows many individuals to congregate at a food source; like the open-country weaver birds, these primates live in groups. In a large group, males compete with each other for mating rights and hence large male body size has been selected for.

Table 2.2 Crook and Gartlan's (1966) division of the primates into five 'adaptive grades'

	Exemplary species	Habitat	Diet	Activity	Group size	Reproductive unit	Sexual dimorphism
Grade I	Galago Lepilemur Microcebus	Forest	Insects	Nocturnal	Solitary	Pairs	Slight
Grade II	Indiri Lemur Hylobates	Forest	Fruit or leaves	Crepuscular or diurnal	Very small groups	Single male with family	Slight
Grade III	Colobus Saimiri Gorilla	Forest or forest edge	Fruit or fruit and leaves	Diurnal	Small groups	Multi-male groups	Slight to fairly marked
Grade IV	Macaca Cercopithecus aethiops Pan	Forest edge or tree savannah	Vegetation —omnivore	Diurnal	Medium to large groups	Multi-male groups	Marked
Grade V	Erythrocebus patas Papio hamadryas Theropithecus gelada	Grassland or arid savannah	Vegetation —omnivore	Diurnal	Medium to large groups	One male groups	Marked

Crook and Gartlan's approach was to categorise the primates into a small number of discrete groups. This raises two main problems. First of all, variation in features such as home range size and group size is continuous and so division into hard and fast groups is a bit arbitrary. Because the groups are subjectively defined it is difficult for subsequent workers to categorise new species in the scheme. Second, different aspects of social organisation such as breeding system and group size do not necessarily vary together in the same way. For example, two species of primate could have the same breeding system but live in different sized groups.

The most recent attempt to unravel the complexities of primate social organisation has been by Tim Clutton-Brock and Paul Harvey. They have tried to avoid these problems firstly by measuring the various aspects of social behaviour and morphology on a continuous scale. Secondly, they use multivariate statistics to tease out the effects of several different ecological variables on the same traits, and to analyse the influence of ecological factors on each aspect of social organisation independently. Their third improvement in approach is a careful consideration about which taxonomic level should be used for analysis, e.g. species, genus, sub-family or family.

This last problem is one about the independence of data points. Imagine plotting all the species of primates on a graph in order to investigate the relationship between body weight and some interesting variable such as home range size, brain size or mating system (e.g. females per male in a breeding group). On our graph we would find that within a genus all the species will be clumped together in a cluster of points. For example, all six species of gibbons are of similar body weight, all are monogamous, arboreal and eat fruit. Our problem

40

Chapter 2
Ecology and
Adaptation:
Comparison
between
Species

is should we treat these as six independent points or just one point in any statistical analysis? If we treated them as six independent points our analysis may be biased because it would reflect phylogeny rather than ecology; all six gibbons may be descended from a single ancestor which was monogamous, arboreal and ate fruit. Because species within a genus tend to have similar characteristics due to phylogenetic constraints, analysis of species data will be statistically biased by those genera containing large numbers of species.

In picking the taxonomic level for our analysis a good rule to use is to choose, within the order (in this case the primates), the taxonomic level where the maximum variance is shown in the variable we are considering. This level can be identified using a nested analysis of variance. As we go up the taxonomic levels the variance will increase. Variance between species within a genus will be small; between genera within a subfamily it will be greater and between families across the whole order of primates there will be still wider scatter in our points. For the primates it is found that maximum variance is shown in most ecological and morphological variables when different genera are considered as independent points (e.g. all six gibbons would be treated as one point only). No additional variance is shown if families are considered rather than genera. Therefore, using the genus as the level for analysis means that there will be no additional variance in the sample which might obscure the results because it is due to taxonomic relationship rather than ecological differences (Harvey & Mace 1981).

We shall now consider some examples of the comparative approach to primate social organisation and morphology, treating different genera as independent points for analysis, to illustrate how comparison has become a more rigorous and objective exercise.

HOME RANGE SIZE

Larger animals need to eat more food and so, in general, we would expect them to have larger home ranges. Therefore, if we want to examine the influence of an ecological variable, such as diet, on home range size, we have to control for body weight as a confounding variable. When home range size is plotted against the total weight of the group that inhabits it, as expected the larger the group weight the larger the home range (Fig. 2.4).

The influence of diet on home range size can be seen when the specialist feeders (insectivores, frugivores) are separated from the leaf eaters (folivores); the specialist feeders have larger home ranges for a given group weight. The probable explanation is that fruit and insects are more widely dispersed than leaves and so specialist feeders need a larger foraging area in which to find enough food.

These general trends are confirmed by more detailed studies of particular species. The red colobus monkey (*Colobus badius*) is a

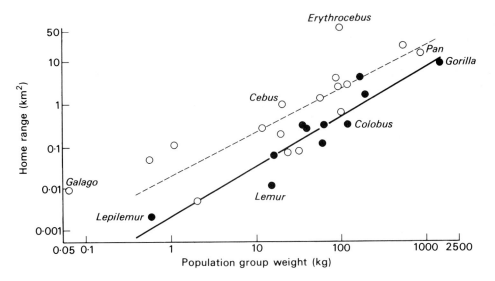

Fig. 2.4 Home range size plotted against the weight of the group that inhabits the home range for different genera of primates. The solid circles (●) are folivores, through which there is a solid regression line. The open circles (○) are specialist feeders (insectivores or frugivores) and the regression line through these points is dashed. Some of the genera are indicated by name (from Clutton-Brock & Harvey 1977).

specialist feeder, eating shoots, fruit and flowers. The food is dispersed in clumps and this species wanders over a large home range of about 70 hectares. The black and white colobus (*C. geureza*) is a generalist, eating leaves of all ages. Its food supply is dense and evenly distributed and its home range is only 15 hectares (Clutton-Brock 1975).

SEXUAL DIMORPHISM IN BODY WEIGHT

In primates, males are often larger than females. Two hypotheses could explain this observation. Sexual dimorphism could enable males and females to exploit different food niches and thus avoid competition (Selander 1972). If this was true, then we might predict that dimorphism would be greatest in monogamous species where male and female usually associate together and feed in the same areas. Alternatively it could have evolved through sexual selection, large body size in males being favoured because this increases success when competing for females (Darwin 1871). If sexual competition is important then we would predict that dimorphism should be greater in polygamous species, where large male size would be especially advantageous because a male could potentially monopolise several females.

The comparative data show no sign of the trend predicted by the niche separation hypothesis but do support the sexual competition hypothesis; the more females per male in the breeding group, the larger the male is in relation to the female (Fig. 2.5).

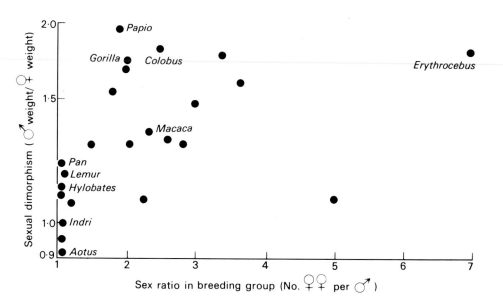

Fig. 2.5 The degree of sexual dimorphism increases with the number of females per male in the breeding group. Each point is a different genus, some of which are indicated by name (from Clutton-Brock & Harvey 1977).

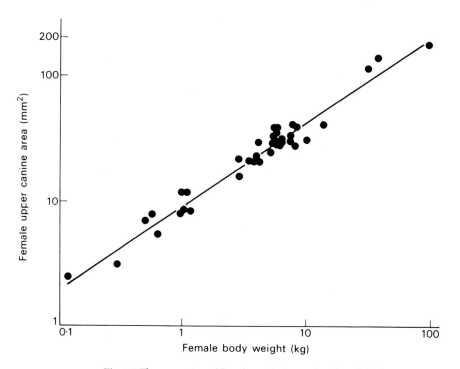

Fig. 2.6 The regression of female tooth size against female body weight. Each point is a genus (from Harvey, Kavanagh & Clutton-Brock 1978). Tooth size increases with body weight so a difference between the sexes in tooth size could just reflect a difference in body size. Body size as a confounding variable can be controlled for by seeing whether a male's tooth size is greater than expected for a female of the same body weight.

*Chapter 2
Ecology and
Adaptation:
Comparison
between
Species*

Males often have larger teeth than females. Again, two hypotheses can be suggested (Harvey, Kavanagh & Clutton-Brock 1978). Large teeth may have evolved in males for defence of the group against predators. Alternatively males may have larger teeth for competition with other males over access to females. There is the problem here of body weight as a confounding variable; males are larger than females and so a difference between the sexes in tooth size could just reflect a difference in body size.

This can be controlled for by calculating the line of best fit when female tooth size is plotted against body weight. If the tooth size of a male is now plotted on the same graph, it can be seen whether its size is greater than expected for a female of the same body weight (Fig. 2.6). The results show that in monogamous species male tooth size is as expected for a female of equivalent body weight. However it is larger than expected in harem forming species. These data support the sexual competition hypothesis for the evolution of larger teeth in males. However we cannot exclude the predator defence hypothesis because maybe the harem forming species are the ones most vulnerable to predation.

The analysis can be taken a step further by considering species where several males live together in a group (multi-male troops). It is found that, within this type of social organisation, the males of terrestrial species have larger teeth for their body size than arboreal species. Therefore even within the same mating system there is a difference in tooth size in different habitats. The terrestrial environment is usually thought to present greater risks of predation and so predator pressure may have been responsible for the evolution of larger teeth in terrestrial species.

Our conclusion is that both sexual competition and predation may have influenced the evolution of sexual dimorphism in tooth size. There is also the further possibility that differences in tooth size are important in reducing diet overlap between the sexes and so preventing competition for food. This example shows that, even with careful analysis, it is difficult to tease out the effect of several variables on the evolution of a trait.

BRAIN SIZE

Brain size increases with body size, so once again we must control for this effect before the ecological correlates of brain size can be examined. Clutton-Brock & Harvey (1980) considered seven families of primates and for each genus within a family they plotted average brain weight against average body weight. For example, within the Cercopithecidae they compared *Macaca* (rhesus), *Papio* (baboons), *Mandrillus* and *Colobus* monkeys. When the brain weight of a genus is plotted against body weight, the deviation from the line of best fit gives

44

Chapter 2
Ecology and
Adaptation:
Comparison
between
Species

a measure of brain size relative to body weight or 'comparative brain size' for the genus in a particular family (Fig. 2.7a). When comparative brain size is plotted against percent foliage in the diet, the interesting result to emerge for all seven families is that leaf eaters have smaller brains than fruit eaters (Fig. 2.7b). The distribution of leaves is denser and more predictable than that of fruit and so leaf-eaters may have smaller brains because they do not need such a large memory store to exploit their food efficiently. However perhaps we could also turn the argument the other way round. Leaf eaters need an enormous digestive system to process their food and a well-fed male colobus monkey may even be difficult to tell apart from a pregnant female! Therefore the brain size of a leaf eater may be small relative to its body weight because selection has favoured a large body weight.

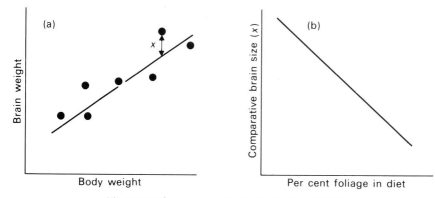

Fig. 2.7 (a) Shows schematically that brain weight increases with body weight. Each point represents a genus within a particular family. The deviation from the regression for a genus (x) gives a measure of its comparative brain size. When this is plotted against per cent foliage in the diet, (b), the regression shows that genera with more foliage in the diet have comparatively small brains for their body weight.

DIFFERENT KINDS OF PRIMATE GROUPS

Both Crook and Gartlan and Clutton-Brock and Harvey considered group *size* as a variable and sought for correlations with ecology. The general trend is for ground-dwelling diurnal species to live in larger groups than nocturnal arboreal species (Table 2.2). The nocturnal species may rely on crypsis in the tree canopy as a means of decreasing predation. Diurnal primates however, especially those living on the ground, are very conspicuous and they may live in large groups for safety against predators (see chapter 4).

Richard Wrangham (1980) has emphasised that simply considering group size misses the important point that there are different *kinds* of primate groups. For example, in species like the Hanuman langur (*Presbytis entellus*) and the olive baboon (*Papio anubis*), the group consists of resident, closely related females plus immigrant males. The females form permanent social bonds and spend a lot of time feeding together and grooming each other. In other species, like the chimpan-

45

Chapter 2
Ecology and
Adaptation:
Comparison
between
Species

zee (*Pan troglodytes*) and the mountain gorilla (*Pan gorilla berengei*) there are no such strong relationships between the females and they frequently move between groups (Fig. 2.8). Can we explain these different *kinds* of groups by differences in ecology?

Wrangham argues that group living will have different costs and benefits for each sex. For males the key resource, namely that which limits reproductive success, is fertile females. For females, on the other hand, the key resource is food (see chapter 6 for why there may be sex differences in key resources). A comparison across primate species shows that groups with strong female bonds occur where the species is exploiting food supplies distributed in discrete, defensible patches. It might pay the females to cooperate with each other to defend these patchy food supplies (e.g. fruit-laden trees) against other groups. By co-operative defence each individual female is presumed to do better than it would by feeding alone.

What about the other kind of groups where female relationships are much looser? It turns out that these species feed on food supplies, such as leaves and stems, which tend to be more evenly distributed throughout the habitat (e.g. mountain gorilla). Such resources are presumed to be less defensible and a female may have no need to form a strong bond with others to gain access to food. Strong bonds are also not formed when the females feed on very small patches of high quality food such as ripe fruit (e.g. chimpanzees). Here, the females may not be able to travel together in permanent close-knit groups because they would interfere with each others' foraging efficiency.

The reader will have noted that these arguments are rather like those advanced at the beginning of the chapter to 'explain' social organisation in the weaver birds. The explanations are plausible, but they have been invented to fit the observations. As such they should be treated as a tentative hypothesis to be tested with rigorous data. Field observations and experiments are needed to test whether it really does pay females of some species to cooperate over food defence. Measurements are needed to see whether some food supplies really are more 'patchily distributed' than others. However, Wrangham's hypothesis raises the fascinating possibility that ecological variables not only influence group size but also the kind of relationships within a group.

The comparative approach reviewed

The statistical approach we have described for the primates is certainly a major improvement on the first applications of the comparative method. To summarise, the main improvements are:

1 Different aspects of social organisation are treated independently and as continuous variables

2 Confounding variables are dealt with in a rigorous manner. This enables us to avoid the pitfall of Victorian sexists who took delight in pointing out that a man's brain is larger than a woman's and

(a)   (b)

Fig. 2.8 Two photographs to illustrate different kinds of primate groups. In chimpanzee (*Pan troglodytes*) groups, (a), the females do not form very strong relationships but in vervet monkeys (*Cercopithecus aethiops*) they do, (b), spending a lot of their time grooming each other. Richard Wrangham has suggested that ecological variables may influence the evolution of relationships within a primate group (photos by Richard Wrangham and Phyllis Lee).

concluded that men must therefore be more intelligent. They ignored an important confounding variable; for a given body weight there is no difference in brain size between men and women.

3 Care is taken to choose the most appropriate taxonomic level for analysis.

4 The data are used wherever possible to discriminate between alternative hypotheses such as predation or sexual competition.

The end result of many of the analyses is a plausible interpretation which may be treated as an hypothesis for further testing. In conclusion, the comparative approach is very useful for looking at broad trends in evolution and the general relationship between social organisation and ecology. It generates hypotheses which can be used as predictions for other groups of animals. It can also be used to test hypotheses which are not amenable to experimentation, such as the effect of polygamy on sexual dimorphism. Furthermore, it is impressive in the way it shows how diet, predation, social behaviour and body size, for example, can all be interrelated.

However, we will need a different approach to understand in detail why animals adopt particular strategies in relation to their ecology. Can we actually measure patch structure and predation risk and then come up with precise predictions as to how an animal will behave? Can we explain why a monkey goes round in a group of 20 rather than one of 16 or 25, why its home range is 10 hectares rather than 8 or 12 hectares, and why it spends 1 hour in a patch of fruit? Indeed we can attempt to answer precise questions like these. No-one has yet attempted this for anything as complicated as primate social behaviour. However a start has been made for simpler kinds of behaviour using optimality theory and an experimental approach. This will form the basis of the next chapter.

47

Chapter 2
Ecology and
Adaptation:
Comparison
between
Species

Summary

The influence of ecology on the evolution of social organisation can be studied by comparing different species and seeing whether differences in behaviour are correlated with differences in ecology. In weaver birds, antelopes and primates the main ecological factors determining the evolution of behaviour are the distribution and abundance of food and predators. These have been shown to influence group size, the relationships between individuals within a group, home range size, mating behaviour, sexual dimorphism and brain size. Two of the main problems in comparative studies are those of confounding variables and the choice of taxonomic level for comparison.

Further reading

Clutton-Brock & Harvey (1979) provide a good discussion of some of the problems in using the comparative approach. David Lack's (1968) book applies the comparative method to the breeding biology of birds.

In this chapter we have not had the space to go into details on any one species of primate. Good accounts of particular species are those of the Altmanns (1970) and the Dunbars (1975) on baboons, Goodall (1968) on chimpanzees, Hrdy (1977) on langurs, and Struhsaker (1975) on the red colobus monkey.

Chapter 3: Economic Decisions and the Individual

Experimental studies of adaptation

In the last chapter we showed how comparisons between species can be used to study adaptation, by correlating differences between species in behaviour with differences in diet or habitat. The comparative approach is impressive because it reveals that a whole variety of features such as body size, brain weight, group size, and home range area can all be related to diet and the impact of predators. But we also saw that one of its limitations is that it produces qualitative rather than quantitative interpretations of how behaviour is adapted to the environment. We now turn to a quite different, and complementary, way of looking at how selection moulds behaviour. Instead of broad scale comparisons between species, the emphasis will be on the behaviour of individuals of the same species and analysing their behaviour in terms of *costs* and *benefits*.

The idea of trying to measure costs and benefits grew out of Niko Tinbergen's experimental approach to studying the survival value of behaviour. For example, Tinbergen observed that in a colony of black-headed gulls (*Larus ridibundus*) nesting on sand dunes in Northwestern England, incubating parents always pick up the broken eggshell after a chick has hatched and carry it away from the nest (Fig. 3.1). Although carrying the shell takes only a few minutes each year it is crucial for the survival of the young. The eggs and young of the blackheaded gull are well camouflaged against the grass, sand and twigs around the nest. The inside of the broken shell, however, is white and highly conspicuous. Tinbergen carried out an experiment to test the hypothesis that the conspicuous white broken shell reduces the camouflage of the nest. He painted hens' eggs to resemble cryptic gull eggs and laid them out at regular intervals in the gull colony. Next to some he placed a broken shell. The results confirmed his prediction that the cryptic eggs would be much more likely to be discovered and eaten by predators such as crows if they were close to a broken shell. So it is easy to visualise why the parent benefits by removing the conspicuous empty shell soon after the chick has hatched: the camouflage of the brood is preserved and the likelihood of the parent perpetuating its genes is increased. But there is more to the story than this. The parent does not remove the eggshell immediately; it stays with the newly hatched chick for an hour or more and then goes off with the shell. In order to explain the delay in removing the shell we have to introduce the idea of a trade-off between costs and benefits. If

Fig. 3.1 A black-headed gull removing an eggshell from its nest. Photo by N. Tinbergen.

the parent flies off with the shell at once, it has to leave the newly hatched chick unattended (the second parent is away at the feeding grounds fuelling up for its next stint at the nest). Tinbergen observed that the new chick, with its plumage still wet and matted, is easily swallowed and therefore makes a tempting meal for a cannibalistic neighbouring adult, and a gull never turns down the chance of a good meal. However, when the chick's down has dried out and become fluffy it is much harder for a gull to swallow, and is therefore less vulnerable to attacks from neighbours. The parent's delay before removing the shell therefore probably reflects a balance between the benefits of maintaining the camouflage of the brood and the costs associated with leaving a newly hatched chick at its most vulnerable moment.

When the balance between costs and benefits is changed, the length of the parent's delay might also be expected to change. This is borne out by observations of the oystercatcher, another ground nesting bird with camouflaged eggs and young. The oystercatcher (*Haematopus ostralegus*) is a solitary nester and cannibalism by neighbours is therefore not a risk associated with leaving the newly hatched chicks.

The parents benefit by restoring camouflage of the nest as soon as possible after hatching and as expected the parent removes broken eggshells more or less as soon as a chick has hatched and before its down is dry.

OPTIMALITY MODELS

Tinbergen's study of eggshell removal illustrates how experimental studies of costs and benefits can be used to unravel behavioural adaptations, but it has an important limitation. The hypothesis about the trade-off between camouflage and chick vulnerability made only a qualitative prediction. The idea would be consistent with observations of gulls removing eggshells 1, 2, 3 or perhaps even 4 h after the chick has hatched, so that it is hard to test whether the hypothesis is right or wrong. One way of trying to make an hypothesis more easily testable is to try to generate quantitative predictions. If one could predict that the parent gull should remove its eggshell after 73.5 mins then one would have produced a very testable model indeed. This is an approach which has been developed by using *optimality models* to study adaptations. An optimality model seeks to predict which particular trade-off between costs and benefits will give the maximum net benefit to the individual. Thinking back to the gulls, if one could measure exactly how much the survival of the brood is reduced by the conspicuous broken eggshell next to the nest and exactly how the risk of cannibalism by neighbours changes with time since the chick hatched, one could start to calculate the optimum time for the parent to delay removal of the shell. In this case the optimum might well be defined as the time that maximises total reproductive success for the season. But the currency of an optimality model does not have to be survival or production of young. The overall success of an individual at passing on its genes may depend on finding enough food, choosing a good place to rest, attracting many mates, and so on. In solving any of these problems an animal makes decisions, and the decisions can be analysed in terms of an optimal tradeoff between appropriate costs and benefits. For a foraging animal, for example, currencies might be energy and time.

Gathering food

CROWS AND WHELKS

On the West coast of Canada, as in many coastal areas, crows feed on shellfish. They hunt for whelks at low tide, and having found one they carry it to a nearby rock, hover and drop it from the air to smash the shell on the rock and expose the meat inside. Reto Zach (1979) observed the behaviour of Northwestern crows in detail and he noted that they take only the largest whelks and always drop the shell from a height of about 5 m. Zach carried out experiments in which he dropped whelks of different sizes from various heights. This, together with data on the energetic costs of flying and searching gave him the information to carry out calculations of the costs and benefits associated with foraging. The benefit obtained by the crow and the cost paid could both be measured in calories, and Zach's calculations revealed that only the largest whelks (which contain the most calories and break open most readily) give enough energy for the crow to make a net profit while foraging. As predicted from these calculations, the crows ignored all but the very largest whelks even when different sizes were laid out in a dish on the beach.

Usually the crow has to drop each whelk twice or more in order to break it open. Since ascending flight is very costly, Zach thought that the crow might have chosen the dropping height which would minimise the total expenditure of energy in upward flight. If each drop is made from close to the ground, a very large number of drops is required to break open the shell, while at greater and greater heights the shell becomes more and more likely to break open on the first drop (Fig. 3.2a). The experiment of dropping shells from different heights allowed Zach to calculate the total vertical flight needed to break an average shell from different dropping heights (Fig. 3.2b). The dropping height that minimises the total vertical flight is impressively close to the crows' average of 5.2 m. However, the crow would have to expend almost the same total energy in upward flight even if each drop was made from a height somewhat greater than 5.2 m (this is indicated by the very shallow U-shaped curve of Fig. 3.2b) because slightly fewer drops would be needed. Zach suggests that there may be an additional penalty for dropping from too great a height: the whelk may bounce away and be lost from view or may break into so many fragments that the pieces are too small to retrieve.

The story of crows and whelks shows how calculations of costs and benefits can be used to produce a quantitative prediction. The crow seems to be programmed to choose the size of whelk and the dropping height that maximises its net energetic profit, and energy is therefore an important currency for a foraging crow.

BIG AND SMALL PREY

Crows are not the only animals to prefer the most profitable size of prey. Bluegill sunfish, sticklebacks, bumblebees, starfish and great tits are among the species that have been shown to exhibit a similar preference. It is not always the largest prey which are most profitable, because sometimes a large item takes so long to handle (subdue and devour) that two slightly smaller prey could have been eaten in the same length of time. In other words a measure of profitability should take into account not only the net energetic yield of a prey, but also the

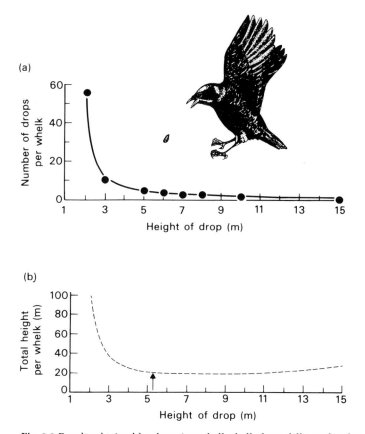

(a)

(b)

Fig. 3.2 Results obtained by dropping whelk shells from different heights (a) fewer drops are needed to break the shell if it is dropped from a greater height. (b) The total upward flight needed to break a shell (no. of drops × height of each drop) is minimal at the height most commonly used by the crows (shown by arrow) (Zach 1979).

time needed to get the yield. Time is usually valuable for animals and achieving the maximum possible *rate* of energy intake may be crucial for survival.

When shore crabs are given a choice of different sized mussels they prefer the size which gives them the highest rate of energy return (Fig. 3.3). Very large mussels take so long for the crab to crack open in its chelae that they are less profitable in terms of energy yield per unit breaking time (E/h) than the preferred, intermediate-sized, shells. Very small mussels are easy to crack open, but contain so little flesh that they are hardly worth the trouble. However the story cannot be as simple as this, because the crabs eat a range of sizes centred around the most profitable ones. Why should they sometimes eat smaller and larger mussels? One possible hypothesis to explain why several sizes are eaten is that the time taken to search for the most profitable sizes influences the choice. If it takes a long time to find a profitable mussel, the crab might be able to obtain a higher overall rate of energy intake by eating some of the less profitable sizes.

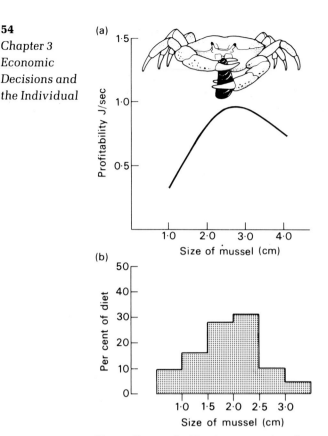

(a)

(b)

Fig. 3.3 Shore crabs (*Carcinus maenas*) prefer to eat the size of mussel which gives the highest rate of energy return. (a) The curve shows the calorie yield per second of time used by the crab in breaking open the shell and (b) the histogram shows the sizes eaten by crabs when offered a choice of equal numbers of each size in an aquarium (Elner & Hughes 1978).

In order to calculate exactly how many different sizes should be eaten we need to develop a more precise argument based on handling time, searching time, and the energy values of the various prey (Box 3.1). The equations in Box 3.1 show the following for the simple example of a predator faced with a choice of two sizes of prey. First, when the more profitable type (higher E/h) is very abundant the predator should specialise on this alone. This is intuitively obvious: if something giving a high rate of return is readily available, an efficient predator should not bother with less profitable items. Secondly the availability of the less profitable prey should have no effect on the decision to specialise on the better prey (this is because the term λ_2 does not appear in equation (5) in Box 3.1). This also makes sense. If good prey are encountered sufficiently often to make it worthwhile to ignore the bad ones, it is never worth taking time out to handle a bad prey regardless of how common they are. The third conclusion from Box 3.1 is that as the availability of the good prey increases there should be a sudden change from no preference (the predator eats both

Box 3.1 A model of choice between big and small prey.
Consider a predator searching for T_s seconds and encountering
two types of prey at rates λ_1 and λ_2 prey per sec. The two kinds
of prey contain E_1 and E_2 calories and take h_1 and h_2 seconds to
handle. They therefore have profitabilities E_1/h_1 and E_2/h_2.

If the predator eats both kinds of prey it will obtain the
following in T_s seconds:

$$E = T_s(\lambda_1E_1 + \lambda_2E_2) \tag{1}$$

and this will take a total time:

$$T = T_s + T_s(\lambda_1h_1 + \lambda_2h_2) \tag{2}$$
$$= \text{searching time} + \text{handling time.}$$

Therefore the predator's rate of intake is (1) divided by (2):

$$E/T = \frac{\lambda_1E_1 + \lambda_2E_2}{1 + \lambda_1h_1 + \lambda_2h_2} \tag{3}$$

(note that T_s has cancelled out).

Let prey type 1 be the more profitable type. If the predator is
to maximise E/T it should specialise on type 1 if:

$$\frac{\lambda_1E_1}{1 + \lambda_1h_1} > \frac{\lambda_1E_1 + \lambda_2E_2}{1 + \lambda_1h_1 + \lambda_2h_2} \tag{4}$$

(energy gain from just eating type 1 > energy gain from eating
both).
This can be rearranged to give:

$$\frac{1}{\lambda_1} < \frac{E_1}{E_2}h_2 - h_1 \tag{5}$$

(note that λ_2 has cancelled out).
This is a useful form of the equation. $1/\lambda_1$ is the expected time
needed to find a type 1 prey.

types when encountered) to complete preference (the predator eats
only the good prey and always ignores the bad ones). Only when the
two sides of equation (5) in the box are exactly equal will it make no
difference to the predator whether it eats one or both types of prey.

An experiment which tried to test these predictions is illustrated in
Fig. 3.4. The predators were small birds (great tits) and the prey were
large and small pieces of mealworm. In order to control precisely the
predator's encounter rate with the large and small worms the experiment involved the unusual step of making the prey move past the
predator rather than vice versa (Fig. 3.4a). The big worms in the
experiment were twice as large as the small ones ($E_1/E_2 = 2$) and h_1 and
h_2 could be accurately measured as the time needed for the bird to pick
up a worm and eat it. During the experiment the bird's encounter rate

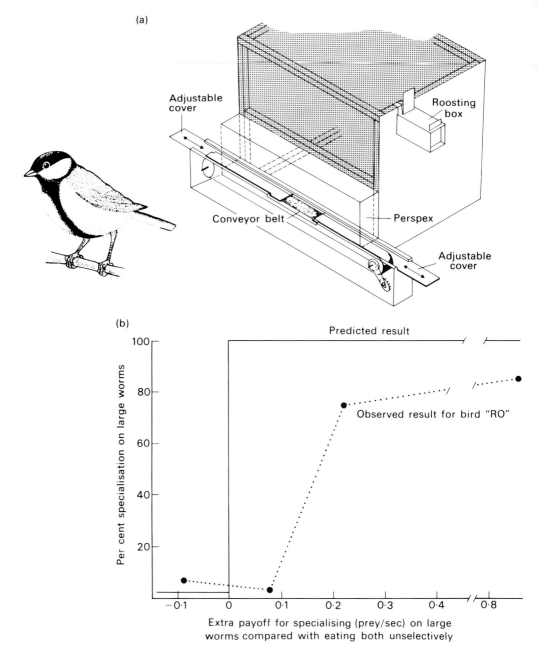

Fig.3.4 (a) The apparatus used to test a model of choice between big and small worms in great tits (*Parus major*). The bird sits in a cage by a long conveyor belt on which the worms pass by. The worms are visible for half a second as they pass a gap in the cover over the top of the belt and the bird makes its choice in this brief period. If it picks up a worm it misses the opportunity to choose ones that go by while it is eating. (b) An example of the results obtained. As the rate of encounter with large worms increases the birds become more selective. The x-axis of the graph is the extra benefit obtained from selective predation. As shown in Box 3.1 the benefit becomes positive at a critical value of $1/\lambda$. The bird becomes more selective about the predicted point, but in contrast with the model's prediction this change is not a step function (Krebs *et al.* 1977).

with large worms was varied so as to cross the predicted threshold from non-selective to selective foraging [equation (5) in Box 3.1]. The results were roughly as predicted, the main difference between observed and expected results was that the switch was not a step but a gradual change (Fig. 3.4b). When big worms were abundant the birds, as predicted, were selective even if small worms were extremely common.

WHAT HAVE WE LEARNED?

The model of choice between big and small worms made precise predictions, so it was easy to test. The discrepancy between the observed and predicted results is important because it can help to lead to a closer understanding of how the animal is designed to make decisions. Perhaps for example, the great tits in the experiments invest a certain amount of effort in sampling the different sized worms and this explains why the switch in Fig. 3.4b was not a step change. By generating a new model, which includes sampling, with new predictions it might be possible to test this idea. The important thing is not just to produce a model that predicts the animal's behaviour, but also to be able to recognise when the hypothesis summarised in the model is incorrect. This is one of the merits of a quantitative model.

In testing an optimality model, we are not trying to test whether an animal is optimal. We start with the *assumption* that natural selection has designed maximally efficient animals (or very close approximations). Ultimately, 'efficient' means good at getting genes into future generations but in order to attain this goal the animal is designed to weigh up costs and benefits at every moment of its life. The aim of an optimality model is to try to gain some insight into the nature of these costs and benefits and how the animal is designed to make decisions (we do not intend to imply here that the animal makes conscious decisions). The model of choice between worms tested the ideas that time and energy are the major components of the cost-benefit equation, that handling time is a constraint, and that animals are designed to maximise their net rate of energy intake. This hypothesis was quite good at explaining the observed behaviour, but not quite right. The problem now is to decide why the model was wrong. Some possible reasons are as follows   *Model failure*

1   One has already been mentioned: perhaps there are some *constraints or limits on the animal's performance which we have ignored.* One constraint, to which we have referred, is the need for the animal to sample the environment; another one might be the need for the animal to eat a certain balance of different nutrients: fat, protein and so on. This second constraint was not important in the experiment with great tits (since large and small worms contained the same nutrients) but it could be vital for herbivores, as we shall describe later.

2   A second reason for the model's failure might be that animals are *not designed to maximise rate of food intake.* Perhaps minimising variation in rate of food intake, or minimising risk of predator attack while feeding could be a more important goal than maximising rate of intake for the animals survival.

3   The hypotheses about constraints and the goal might both be correct, but there may be components of the cost–benefit equation that have not been measured (e.g. handling large prey might be relatively more costly in terms of energy).

4   Finally, the basic assumption underlying the model, that evolution has produced maximally efficient design, could be wrong. It could be incorrect because, for example, the environment might change rapidly enough for animals to be ill-adapted to present conditions. This should be an argument of last resort since unlike the other three suggestions, it does not immediately generate new, testable hypotheses and experiments. It is more fruitful to explore the first ideas such as constraints and alternative optimality criteria.

NUTRIENT CONSTRAINTS: MOOSE AND PLANTS

As a general rule nutrient quality of food is more important to herbivores than to carnivores and insectivores. This is because plants often lack essential dietary components and only by careful selection of plant species can a herbivore obtain a balanced intake. For example, the diet of moose (*Alces alces*) on the shores of Lake Superior in Michigan is strongly influenced by sodium requirements. The moose feed in two habitats: forest, where they browse on deciduous leaves, and small lakes, where they crop plants growing under water. The aquatic plants are rich in sodium but relatively poor in energy, while terrestrial plants have little sodium but a high energy content. The moose need both energy and sodium to survive and therefore have to eat a mixed diet, but to predict the exact mixture involves making an optimality model.

Since the diet of the moose contains two components we can plot it as a point on a graph, the axes of which are intake of terrestrial and aquatic plants (Fig. 3.5). If, for example, the moose ate largely terrestrial plants with just an occasional aquatic, its diet would be represented by a point at the lower right-hand side of the graph. Now as we have already said, the daily diet has to contain a certain minimum amount of sodium. This is represented in the graph as a constraint line: the horizontal dot-dash line shows the minimum intake of aquatic plants needed to satisfy sodium requirements. But this is not the only constraint on the animal's diet. It also needs a certain amount of energy per day. This could be obtained by eating a pure diet of y grams of aquatic plants or x grams of terrestrial plants or a mixture of the two, as shown by the solid line in Fig. 3.5. This line shows the mixture of plants that would provide just enough energy to

survive the day. Finally, the diet is constrained by the size of the
moose's rumen. The moose has a specially modified stomach, the
rumen, in which food is slowly fermented by micro-organisms prior to
digestion by the moose. The size of the rumen sets an upper limit on
the amount of food that can be processed at any one time, and
therefore limits the total daily intake. The broken line on Fig. 3.5
shows the maximum amount of food that could be eaten per day with
different combinations of terrestrial and aquatic plants.

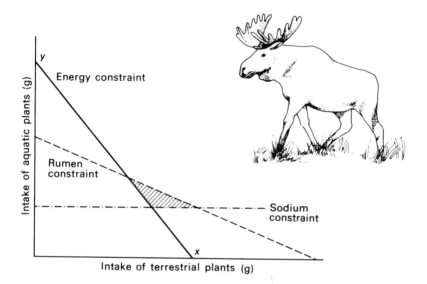

Fig. 3.5 The diet of moose is constrained by the requirements for sodium and energy:
the daily requirements are shown as the dot-dash and solid lines respectively and
the moose has to eat a mixture of plants which lies in the space above these two
lines. The third constraint is the size of the moose's rumen (broken line). Aquatic
plants are bulkier than terrestrial ones so fewer grams of these can be fitted into the
rumen. The mooses' diet was found to lie, as predicted, inside the region of the
shaded triangle (Belovsky 1978).

The total effect of these constraints can now be assessed. Only the
diets inside the small shaded triangle on Fig. 3.5 satisfy all three
constraints. The diet has to be above the sodium line, above the energy
line, and below the rumen line. But where inside the triangle is the
optimal diet? This depends on the goal or optimality criterion. If the
moose is designed, for example, to maximise its daily sodium intake,
the diet should include as much aquatic material as possible and lie in
the top left corner of the triangle. If the moose was designed to
minimise the time spent in water each day, its diet might be near the
bottom right of the triangle. Gary Belovsky (1978) carried out a
detailed study of the diet and found that the mixture of plants eaten
was at the point within the triangle which would be predicted if the
moose maximises its daily energy intake subject to the constraints of
sodium need and rumen size.

Imagine you are offered the choice of two daily food rations: one is fixed at 10 sausages per day, the other is uncertain; on half the days you get 5 sausages and on the other half, 20 sausages. Although the *average* of the second diet is higher than that of the first, it is a riskier option since there is no way of telling whether you will get 5 or 20 on any particular day. Which is the better option? The answer depends on the benefit (or 'utility' in economic jargon) of eating different numbers of sausages per day. If a diet of 10 is enough to survive on while 5 is not, then little is to be gained by choosing the risky option. If, on the other hand, 10 is not quite enough to survive on, the only viable option may be to take the risk and hope for 20 sausages. This option offers a 50 percent chance of survival while the certain option offers no chance.

This idea has been tested in an ingeneous experiment by Tom Caraco *et al.* (1980). They offered yellow-eyed juncos (*Junco phaeonotus*) (small birds) in an aviary a choice between two feeding options: either a risky one or one with a certain pay-off. Both feeding stations offered the same mean reward rate, but at one the reward was always the same (e.g. two seeds) while at the other the pay-off was variable (e.g. half the time no seeds, half the time four seeds). They tested the birds after two lengths of starvation period: after the short starvation period of one hour the certain option provided just enough food for daily survival, but after a period four hours' starvation the birds had to take the risk of the uncertain choice to stand a chance of getting enough to eat. As predicted by the argument above, the birds changed their preference from the certain to the risky option as the starvation period increased. The conclusion is that variation in rate of food intake may sometimes have a more important effect than the average rate.

Copulating dungflies

The optimality approach used to analyse foraging decisions can equally well be applied to other kinds of behaviour. Copulating dungflies will serve as an example.

A fresh cowpat in an English meadow is soon invaded by a swarm of yellow dungflies. The first flies to arrive are males seeking out females to mate with; the females start to arrive a few minutes later and they are 'captured' by males as soon as they arrive on or near the pat. Immediately after copulation the females lay their eggs on the dungpat where the larvae hatch out and grow. Males compete vigorously for females and often one will succeed in kicking a rival off a female during copulation and will take her over. When two males mate with the same female the second one is the individual whose sperm fertilises most of the eggs. Geoff Parker (1978) showed this by the clever technique of irradiating males with cobalt[60], which sterilises them but does not alter sperm activity (the sperm can still fertilise an egg but the egg does not develop). If a normal male is allowed to mate after a sterile one about 80 percent of the eggs hatch, whereas if the sterile male mates second only 20 percent of them hatch. The conclusion from these 'sperm competition' experiments is clear: the second male's sperm fertilises about 80 percent of the eggs. It is not surprising, therefore, that after a male has copulated he sits on top of the female and guards her until the eggs are laid, only relinquishing his position to a rival male after a severe struggle.

When a second male takes over (or when a male encounters a virgin) how long should he spend copulating? Parker carried out sperm competition experiments in which he interrupted the second male's copulation after different times; this showed that the longer the second male mates the more eggs he fertilises, but the returns for extra copulation time diminish rapidly (Fig. 3.6a). At first sight it may seem as though the male should copulate for about 100 min, long enough to fertilise all the eggs. But there is a cost associated with a long copulation: the male misses the chance to go and search for a new female. After the male has copulated for long enough to fertilise about 80 percent of the eggs the returns for further copulating are rather small and the male might do better by searching elsewhere for a new mate.

The dungfly's dilemma of how long to spend copulating can be solved graphically (Fig. 3.6b). The graph shows time along the bottom and proportion of eggs fertilised on the vertical axis. Consider a male which has just finished copulating with a female. Before he can copulate with a new female he must first guard his present female until she has laid all her eggs and then he must fly off to search for a new mate. This takes on average 156 min. Once he has found a new female the proportion of eggs fertilised as a function of copulation time is set by the curve measured in the sperm competition experiments. The male cannot vary the guarding time, search time or fertilisation curve but he can choose a copulation time which maximises the proportion of eggs fertilised per min. If the male stays too long with one female he misses opportunities elsewhere, if he stays too short a time he does too much searching and not enough copulating. The best compromise is to copulate for 41 min. This conclusion is reached by drawing a line from point A of Fig. 3.6b to just touch the fertilisation curve. The triangle created by this line has 'time' as its base and 'proportion of eggs fertilised' as its vertical side. The slope of the triangle's hypotenuse is the maximum possible proportion of eggs fertilised per unit time. The average copulation time in the field is 36 min, quite close to the predicted 41 min.

The solution to the dungfly's dilemma is also relevant to other decision problems. For example many animals feed on prey which occur in clumps or patches, and the problem of how long to spend on a patch is exactly analogous to the dungfly problem (Fig. 3.7; Box 3.2).

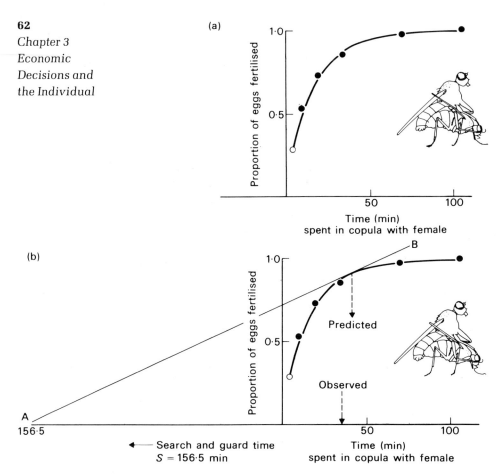

Fig. 3.6 (a) The proportion of eggs fertilised by a male dungfly (*Scatophaga stercoriaria*) as a function of copulation time: results from sperm competition experiments. (b) The optimal copulation time (that which maximises the proportion of eggs fertilised per minute), given the shape of the fertilisation curve and the fact that it takes 156 min to search for and guard a female, is 41 min. The optimal time is found by drawing the line AB (Parker 1978).

Box 3.2

The model of optimal copulation time drawn in Fig. 3.6b is applicable to any example of an animal exploiting patches of resource where patches are depleted by the animal. As a result of depletion, the returns for staying longer in a patch gradually diminish and eventually it pays to move on. The resource could be females (as in the dungfly), food, water and so on.

If the time taken to travel between patches increases, it will pay the animal to spend longer in each patch before leaving as shown in the diagram below. For short travel times, the optimal patch time is T_S and for long travel times it is T_L. The method of determining the optimal time is the same as used in Fig. 3.6.

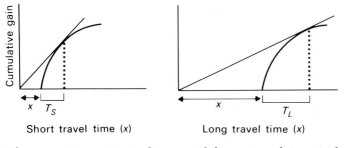

Short travel time (x)   Long travel time (x)

The reason it pays to stay longer with longer travel times is that
the expected returns for moving are lower since the move
includes a longer journey. This prediction was tested by
Richard Cowie (1977) in an experiment using captive great tits
which searched for worms hidden in patches (sawdust-filled
cups). When the travel time was increased the birds spent
longer in each patch before leaving, just as predicted by the
graphs above.

What happens if the average quality of patches changes? The
quality of a patch is represented in the graphs by the gain
curve: good patches have a steeply rising curve and bad
patches have a shallow curve. If all the patches are good, the
expected gain from moving is high and it will pay the animal to
give up and move on quite quickly. If the patches are all poor, it
will pay to persist when the gain rate is lower in each patch
(diagram below), since the returns for moving are low. This
prediction was tested in a study of bumblebees described in
chapter 12.

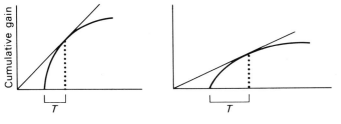

Good patches   Bad patches

The same model can also be used to analyse the problem of
'central place foraging' in which an animal makes repeated
trips to feeding sites and each time brings back a load of food to
a central place such as a nest or food store. Suppose the animal
catches several items on one trip and that the efficiency of
capturing successive items declines because of the hindering
effect of prey already caught and held in the mouth. The result
will be a curve of diminishing returns like that of Fig. 3.6b on
each foraging trip. The graphical method used for dungfly
copulation times can be used to predict the optimal load size
for a given average travel time. The travel time in this case is

the time taken to complete the round trip from the central place to the feeding site (Orians & Pearson 1979). There have been no quantitative tests of this prediction to date, but Kramer & Nowell (1980) showed that chipmunks (*Tamias striatus*) collecting sunflower seeds follow a curve of diminishing returns as they load their cheek pouches, and Orians (1980) found that Brewer's blackbirds (*Euphagus cyanocephalus*) brought larger loads to the nest when they returned from distant feeding sites than from close sites.

(a) (b)

Fig. 3.7 (a) A male dungfly copulating. Courtesy of Dr G. A. Parker and *Behaviour*. (b) The same model as was used to predict the dungfly's copulation time could be used to predict the behaviour of a predator feeding on patchily distributed food such as these goldfinches (*Carduelis carduelis*) feeding on teasel heads. Reproduced by permission from *Nature in the Wild* published by Country Life.

Feeding and predation: a trade-off

When house sparrows feed on a freshly sown field of spring barley they usually stay close to a hedge or bush to which they can rapidly retreat if danger threatens. Even if the availability of grain (and therefore feeding rate) is higher in the middle of the field the sparrows stay at the edge (Barnard 1980). They are not choosing the place with the best feeding rate but instead they seem to adopt a compromise between safety and obtaining enough food. The simplest interpretation is that the sparrow faces two major risks, starvation and capture by a hawk or cat. If the sparrow sat in a bush all day it would be completely safe from hawks but would soon die of starvation. If it fed in the middle of the field it would have no chance of death through lack of food but might soon get eaten by a hawk. Thus the birds' behaviour might be explained by the hypothesis that they feed at a distance from cover which minimises the combined risk of starvation and predators. One prediction of this hypothesis would be that when the weather is very cold and the risk of starvation is more accute, the birds should move further into the field, just as normally shy birds come to a winter feeder near a house in a cold spell.

Manfred Milinski and Rolf Heller (1978, 1979) studied a similar problem with sticklebacks (*Gasterosteus aculeatus*). They placed hungry fish in a small tank and offered them a simultaneous choice of different densities of water fleas, a favourite food. When the fish were very hungry they went for the highest density of prey where the potential feeding rate was high, but when they were less hungry the fish preferred lower densities of prey. Milinski and Heller hypothesised that when the fish feeds in a high density area it has to concentrate hard to pick out water fleas from the swarm darting around in its field of vision, so it is less able to keep watch for predators. A very hungry fish runs a relatively high risk of dying from starvation and so is willing to sacrifice vigilance in order to reduce its food deficit quickly. When the stickleback is not so hungry it places a higher premium on vigilance than on feeding quickly, so it prefers the low density of prey. The balance of costs and benefits shifts from feeding to vigilance as the stickleback becomes less hungry.

Although they have not tested whether sticklebacks are less vigilant when they feed at a higher rate, Milinski and Heller found that predation risk influences choice of feeding rate. When they flew a model kingfisher (*Alecedo atthis*) (a predator on sticklebacks) over a tank containing hungry fish they found that the sticklebacks preferred to attack low rather than high prey densities (Fig. 3.8). This is to be

FEEDING
vs
VIGILANCE

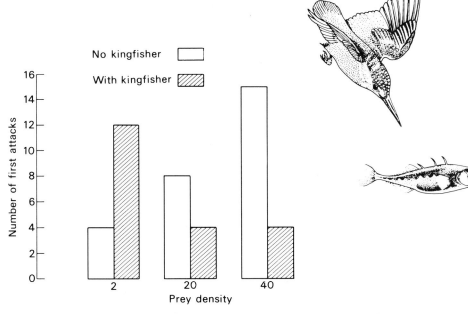

Fig. 3.8 Hungry sticklebacks normally prefer to attack high density areas of prey but after a model kingfisher was flown over the tank they preferred to attack low density areas (Milinski & Heller 1978).

expected if the hungry fish, in spite of its high risk of starvation, places a very high premium on vigilance when a predator is in the vicinity.

An important difference between Milinski and Heller's analysis of foraging and those described earlier is that the cost-benefit calculations include the animal's hunger state. An optimisation model in which the animal's state changes as a result of its behaviour (the fish becomes less hungry as a result of feeding) is referred to as a dynamic, as opposed to static model. In fact the traditional view that an animal's internal state controls its behaviour can be turned on its head and the animal can be seen as using its behavioural repertoire to control the internal state in an optimal way. The influence of the kingfisher on the stickleback is to alter the optimal allocation of time to feeding and vigilance, so that the fish decreases its hunger at a slower rate.

Advantages and limits of optimality models

Let us now summarise the advantages and disadvantages of the optimality approach that we have encountered during this chapter. The three main advantages are.

1 The models make testable and often quantitative predictions: this means that it is relatively easy to see whether the hypothesis is right or wrong.

2 A second advantage is that the assumptions underlying a quantitive model have to be made explicit. The model of choice between big and small worms, for example, had very specific assumptions: prey were recognised instantaneously, encountered sequentially, differed only in energy value and handling time, and so on.

3 Finally, optimality models emphasise the generality of simple decision rules. The model of dungfly copulation could equally well have been tested (and it has) by watching bumblebees collect nectar, or predatory waterboatmen suck the body contents out of mosquito larvae, because these are all examples of an animal exploiting a depleting resource.

The most important limitation of the optimality approach that we have come across so far is the difficulty of telling why there are differences between observed results and predictions. Four possible reasons were mentioned (and there may be more): incorrect assumptions about constraints, inappropriate choice of goal, components of the cost–benefit equation not measured, and poorly adapted animals. There is no simple recipe for distinguishing between these and one has to rely on the intuition of the investigator. A second objection which is often levelled at optimality models is that they say nothing about *mechanisms*. Although animals may behave as if they calculate solutions to cost–benefit equations, no-one seriously suggests that this is how they are programmed to do the job. Instead they probably have simple rules of thumb that give approximately the right answer. It is important to distinguish between the mechanisms (how the animal does it) and the goals (what selection has designed the animal to do). Optimality models only refer to the latter, although they may raise interesting questions about the former. Last but by no means least, we have not considered the possibility that there may be several equally advantageous ways of attaining the same goal. This may be particularly important if the benefit derived from choosing one option depends on the choices made by other individuals in the population. This extra dimension of complexity can still be tackled with a similar approach, as chapter 5 will show. We will return to these points again in chapter 13.

Summary

Behaviour can be viewed as having costs and benefits and animals should be designed by natural selection to maximise net benefit. This idea can be used as a basis for formulating optimality models in which the criterion of maximum benefit, the constraints on the animal, and the currency for measuring benefit are specified. Different kinds of currency might be appropriate for measuring benefits and costs of

Table 3.1
Summary of the decision models described in this chapter. In each example the hypothesis involved proposing both a 'goal' and constraints.

Animal	Decision	Hypothesis	Test
Northwestern crow	Height from which to drop shells to smash them	Height chosen will minimise total ascending flight	Experiments: Drop shells from different heights
Shore crab	Size of mussel to eat	Maximise energy intake per unit handling time	Offer crabs a range of sizes and observe choice
Great tit	How should prey choice change with encounter rate	Maximise energy intake per unit foraging time	Present prey of different sizes at various encounter rates
Moose	Eat aquatic or terrestrial plants	Maximise energy intake per day subject to constraints of sodium requirements	Measure diet
Yellow-eyed Junco	Risky or certain food reward	Risk proneness depends on starvation	Choice of risky and certain food at different levels of starvation
Dungfly	How long to spend copulating	Maximise rate of fertilising eggs	Sperm competition experiments. Measure search time in the field
Stickleback	Feed in high or low density patches of waterfleas	Minimise combined risk of starvation and predation	Vary hunger and predation risk. Observe choice.

different behaviours, for example with feeding behaviour rate of intake might be good currency and with male mating behaviour, rate of fertilising eggs seems reasonable (Table 3.1).

The emphasis of this approach is on quantitative testable predictions. Often the results of experiments deviate slightly from the predictions of simple models; these deviations can be just as valuable as successful predictions in helping to understand how behaviour is designed.

Further reading

Maynard Smith (1978a) is a good review of the problems and advantages of using optimality arguments in ecology, evolution and behaviour, while Lewontin (1979) is critical of the optimality approach. Pyke et al. (1977) is about optimal foraging and the chapters by Krebs and Parker in Krebs and Davies (1978) review work on optimal foraging and dungfly mate searching respectively.

MacFarland (1977) gives a brief review of how motivational (internal) and external costs can be combined to predict optimal trade-offs between different kinds of behaviour.

Chapter 4. Living in groups and defending resources

Show anyone 10 000 flamingos nesting beak by jowl in a colony and the chances are that sooner or later they will ask 'Why on earth are they all nesting so close together?'. In this chapter we will look at why animals live in groups: why flamingos flock, horses herd and sardines shoal. Armed with the methods developed in chapters 2 and 3, comparisons between species and optimality models, we will try to show how ecological forces might favour living in groups.

Comparisons between species suggest that the two main environmental influences on group size are food and predators (chapter 2) and comparisons between populations within a species also emphasise their importance (Fig. 4.1). In many studies either costs or benefits related to feeding and predation have been measured and we will describe some of these in the first part of the chapter, before moving on to consider how different kinds of cost and benefit can be combined to predict optimal group size. Animals which do not live in groups (and also some which do) often defend resources from which they exclude other members of the same species. So the other side of the question 'why live in groups?' is 'when does it pay to defend resources and keep others away instead of joining them?'. This will be discussed in the last part of the chapter.

Living in groups and avoiding predation

The guppies in Fig. 4.1a live in groups when they are in streams where predators are common, which suggests that being in a group might help an individual to avoid becoming a meal. This could happen in several different ways.

INC. VIGILANCE
DILUTION/COVER
GROUP DEFENSE

INCREASED VIGILANCE

For many predators success depends on surprise: if the victim is alerted too soon during an attack, the predator's chance of success is low. This is true, for example, of goshawks hunting for pigeon flocks (Fig. 4.2): the hawks are less successful in attacks on large flocks of pigeons mainly because the birds in a large flock take to the air when the hawk is still some distance away. If each pigeon in the flock occasionally looks up to scan for a hawk, the bigger the flock the more likely it is that one bird will be alert when the hawk looms over the horizon. Once one pigeon takes off the others follow at once.

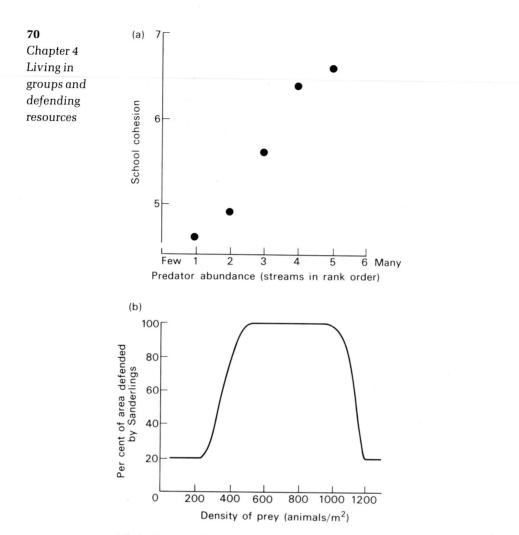

Fig. 4.1 Intraspecific variation in group size may be related to predators and food. (a) Guppies (*Poecilia reticulata*) from different streams in Trinidad: guppies from streams with many predators live in tighter schools than those from streams with few predators. Each dot is a different stream and 'cohesion' was measured by counting the number of fish in grid squares on the bottom of a tank (Seghers 1974). (b) Sanderlings (*Calidris alba*) in Bodega Bay California: The birds defend stretches of beach in some parts of the intertidal zone and feed in roving flocks on other parts of the beach. Whether or not the birds defend territories depends on the density of the major prey, an isopod called *Excirolana linguifrons*. Territories are mainly defended in areas of intermediate prey density. At very low densities there are not enough prey to make defence worthwhile and at very high densities there are so many sanderlings trying to feed that defence would not be feasible because of high intruder pressure. In the area where birds defend territories there is an inverse correlation between territory size and food density (Myers *et al.* 1979).

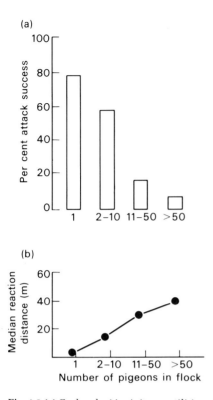

Fig. 4.2 (a) Goshawks (*Accipiter gentilis*) are less successful when they attack larger flocks of wood pigeons (*Columba palumbus*) (b) This is largely because bigger flocks take flight at greater distances from the hawk. The experiments involved releasing a trained hawk from a standard distance (Kenward 1978).

The precise way in which vigilance changes with flock size depends on how individuals in the group spend their time. In ostrich flocks, for example, Brian Bertram (1980) found that each individual spends a smaller proportion of its time scanning than when alone but that the overall vigilance of the group (proportion of time with at least one bird scanning) increases slightly with group size (Fig. 4.3). Therefore each bird in the flock has more time to feed and enjoys greater awareness of approaching lions (a potential predator of ostriches). The increase in vigilance with group size is as predicted if each bird raises its head independently of the others. The ostriches also raise their heads at random time intervals which makes it impossible for a stalking lion to predict how much time it has to creep forward undetected between look-ups by its victim. Any predictable pattern of looking could be exploited by the lion in its tactics of approach.

The problem of how individuals in a group scan is complicated by the fact that in a large group, where overall vigilance is at the maximum value of 100 per cent, it would pay an individual to 'cheat' and spend all its time with its head down feeding. The cheater loses nothing in terms of vigilance because others are busy scanning and it gains extra time to feed. It is not known how this kind of cheating is prevented from evolving, but one suggestion is the following. Although the 'innocent' strategy of scanning regularly regardless of what other's do is susceptible to cheating, a flock made up of more canny individuals, which do not scan unless they have seen their neighbours doing the same thing, might be resistant to cheaters (Pyke *et al.* in prep.). The general point is that even when there is an overall benefit of being in a group, each individual will be expected to try to get more benefit than the others.

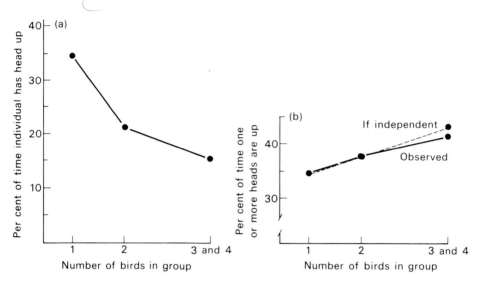

Fig. 4.3 Vigilance in groups. (a) An ostrich (*Struthio camelus*) spends a smaller proportion of its time scanning for predators when it is in a group. (b) The overall vigilance of the group increases slightly with group size, (solid line) as predicted if each individual looks up independently of the others (broken line) (Bertram 1980).

DILUTION AND COVER

Although there is only a slight increase in vigilance with increasing group size in ostriches, the chances that any one individual will be eaten during an attack by lions decreases rapidly with group size, because the lions can kill only one ostrich per successful attack. By living in a group the ostrich dilutes the impact of a successful attack because there is a good chance that another bird will be the victim. To some extent this dilution effect may be offset by the increased number of attacks on larger and more conspicuous groups, but usually the net effect probably favours living in a group, as the following hypothetical example illustrates. An individual antelope in a herd of a hundred has (all things being equal) only a one in a hundred chance of being the victim in a single attack and the herd is not likely to attract more than a hundred times as many attacks as a solitary antelope. In fact if the herd is more vigilant it may pay the predator to concentrate its attacks on small groups and solitary individuals.

One study in which the survival rate of individuals in different sized groups was measured showed an overall benefit of group living from dilution. The monarch butterfly (*Danaius plexippus*) migrates from North America to spend the winter in warmer places such as Mexico. They assemble into enormous communal roosts in which the trees over an area of up to 3.0 ha may be clothed in resting butterflies. The monarch is not a very palatable butterfly, but some birds attack them in the winter roosts. Counts of the remains of predated butterflies showed that predation rate is inversely related to colony size, so the advantage of dilution seems to outweigh any disadvantage of greater conspicuousness in a large roost (Calvert *et al.* 1979).

The dilution effect is probably a very widespread advantage of being in a group and it might explain the strange behaviour of birds such as ostriches and goosanders when they have young. When two females meet, each appears to try and steal the other's young and incorporate them into its own brood. Usually caring for someone else's young doesn't pay, but if predation pressure is severe it might, because of dilution. A more concrete example of the dilution effect comes from a study of semi-wild horses in the Carmargue, a marshy delta in the south of France. In the summer months the horses are plagued by biting Tabanid flies and during this period they are more likely to cluster together in large groups. Measurements of the number of flies per horse in large and small groups showed that horses in a large group are less likely to be attacked. An experiment in which horses were transferred from large to small groups and vice versa confirmed that living in a group gives protection by the dilution effect (Duncan & Vigne 1979).

In some animals dilution is achieved by synchrony in time as well as in space, and this might explain the remarkable 13 and 17 year life cycles of certain species of cicada. These insects live as nymphs underground and the adults emerge after 13 or 17 years depending on the species and location. In the 17 year cicada studied by Dybas & Lloyd (1974) millions of adults (of three species) emerge in synchrony over a wide area, effectively 'flooding the market' so that the chances of any one individual falling victim to a predator is reduced. Lloyd & Dybas (1966) and others have speculated on why the cycle should be 13 or 17 years long and not, for example, 15 or 18. The advantage of a very long dormant stage between emergence periods is that it forces specialist predators and parasites out of business. When there are no cicadas around for 13 or 17 years the predators have either to die or to switch to other prey or to become dormant themselves. The very long cycle must have evolved as a result of an 'evolutionary chase' in which both cicadas and their predators gradually extended their life cycles until the cicadas eventually 'won'. The significance of the 13 and 17 year periods is that these are prime numbers which means that a predator could not regularly fall into synchrony with the cicadas if it had a short life cycle of which the cicada cycle is a multiple. If, for example cicadas had a 15 year cycle, predators with 3 or 5 year life cycles would fall into step with their prey every fifth or third generation.

This idea remains an interesting speculation, but synchrony is certainly an advantage. Field evidence shows that cicadas emerging at the peak of the cycle have a lower chance of succumbing to predators than those emerging early or late (Simon 1979). Selection therefore acts to maintain synchrony once it is established.

Just as a cicada in the middle of the emergence period is safer than one at either end, individuals in the middle of a flock, school or herd may enjoy greater security than those at the edge. If the predators pick off victims from the edge, each member of the group should jockey for a central position and, in effect, seek cover behind the others (Hamilton 1971). This may explain why starling flocks, for example, bunch together in a tight group when a predator approaches. Why should predators attack the edge of the group? The old trick of throwing three tennis balls to a friend at the same time shows how difficult it is to track one of a number of rapidly moving objects in the visual field for long enough to catch it. There is some evidence that predators suffer from the same type of confusion when attacking a dense group of prey (Neill & Cullen 1974) and this may provide an explanation of why attacks should be directed at the edge of a group.

GROUP DEFENCE

Prey animals are often not just passive victims and by living in a group they may be able to defend themselves against the unwelcome attentions of a predator. In colonies of blackheaded gulls nesting pairs will mob a crow when it flies near their nest and in the centre of a dense colony many gulls mob the crow at the same time because it is close to many nests. The effect of this is to reduce the success of the crows in hunting for gulls' eggs (Kruuk 1964) (see also Fig. 4.4a).

COSTS OF BEING IN A GROUP

As we mentioned earlier, one of the costs of group living might be increased conspicuousness. This cost was studied experimentally by Malte Andersson (Andersson & Wicklund 1978) using artificial nests of the fieldfare, a thrush-like bird which breeds colonially in Scandinavian boreal forests. The bulky nests are quite conspicuous and a colony of artificial nests attracted more predators than did solitary nests. However fieldfares vigorously mob crows and other predators and Andersson & Wicklund found that artificial nests placed near a colony of fieldfares survived better than those placed near solitary fieldfare nests. They concluded therefore that the benefit of group mobbing by members of a colony more than offsets the disadvantage of being conspicuous.

(a)

(b)

Fig. 4.4 (a) Group living and avoiding predators. In dense colonies of guillemots (*Uria aalge*) like this one, breeding success is higher than in sparse colonies because of more effective defence against nest predators such as gulls (Birkhead 1977) photo by T. R. Birkhead. (b) Group living and hunting for food: Hyenas (*Crocuta crocuta*) can successfully attack prey which are larger than themselves because they hunt in a group (Kruuk 1972). Photo by Hans Kruuk.

Living in groups and getting food

FINDING GOOD SITES

The comparative studies described in chapter 2 revealed that species which feed on large ephemeral clumps of food such as seeds or fruits often live in groups. For these animals the limiting stage in feeding is the problem of finding a good site: once the patch has been found there is usually plenty of food, at least for a short while. Peter Ward and Amotz Zahavi (1973) developed the idea that communal roosts and nesting colonies of birds may act as 'information centres' in which individuals find out about the location of good feeding sites by following others. The idea is that unsuccessful birds return to the colony or roost and wait for the chance to follow others who have had more success on their last feeding trip. Unsuccessful birds might recognise successful ones by, for example, the speed with which they fly out from the colony on their next trip.

It is perhaps unfortunate that Ward and Zahavi used the phrase 'information centre' which carries with it a connotation of mutual cooperation in the transfer of information, as for example in a honey bee or ant colony. As we shall see in chapter 10 there are special reasons to expect cooperation in social hymenopteran colonies, but these do not apply to bird roosts or nesting colonies. 'Mutual parasitism' might be a more appropriate label here, since the successful foragers are in effect parasitised by unsuccessful birds. Each individual is out to maximise its own success and not the success of the colony as a whole. In some species the 'informer' might be unable to avoid being followed because it is conspicuous when it leaves the nest, for example seabirds leaving a colony on a cliff. The 'informer' might, however, benefit from being followed. The benefit could be long term: on a later trip the leader becomes a follower, or short term: there may be an advantage to feeding in a group because, for example, of the reduced risk of predation. If these benefits do not outweigh the disadvantage of competition at the feeding grounds which arises from being followed, a successful bird should conceal as much as possible information about its success.

The most direct experimental test of Ward and Zahavi's idea is a study of communally roosting weaver birds (*Quelea quelea*) by Peter de Groot (1980). Quelea nest in a colony and roost in groups sometimes estimated to contain over a million birds. They are serious agricultural pests in parts of central Africa and can devastate a grain field in a few hours. De Groot's experiments were done on a somewhat more modest scale (Fig. 4.5). Two groups of birds roosted together in the large aviary labelled 'x' and had access to foraging areas in the small compartments labelled 1–4. The birds could not see into the foraging compartments from the roosting area but had to pass through small entrance funnels to explore them for food or water. In one experiment one of the groups (A) was trained to find water in one of the four compartments and the other group (B) was trained separately to find food in another compartment. The two groups were allowed to roost together and were deprived of food or water. When they were thirsty, the birds of group B followed A to the drinking site, and when they were hungry A followed B to the feeding site. Somehow the 'naive' birds assessed that the other group was knowledgeable and followed it to the resource supply.

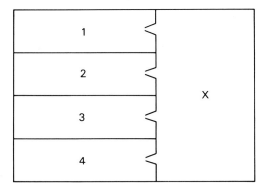

Fig. 4.5 An experiment to test the 'information centre' hypothesis with *Quelea*. The birds roost in the large area labelled X and feed in the smaller compartments labelled 1–4.

In a second experiment group A was trained to forage on a good food supply (pure seed) in one of the compartments while group B was separately trained to fly to another compartment for a poor food supply (seed in a bed of sand). When the two groups roosted together, the members of the second group followed the first group when they left the roost at dawn. It is not yet known how the birds recognise which individuals to follow.

CATCHING DIFFICULT PREY

Individuals in groups may be able to capture prey which are difficult for a single individual to overcome either because the prey is too large for one predator to handle (e.g. lions hunting adult buffalo) or because it is too elusive for one predator to catch (e.g. killer whales hunting porpoise). When the prey themselves are in a group the predators may, by hunting in a group, succeed in separating a victim from its companions and subsequently chasing it until they overtake it. This is how predatory fish such as the jack (*Caranx ignobilis*) hunt for schooling prey. Individuals in a group are more successful than single fish when hunting for schools of Hawaiian anchovy (*Stolephorus purpureus*) (Fig. 4.6). However the benefit is not shared equally between members of the hunting group: fish at the front of the school when it is chasing a prey get more than those at the back (Fig. 4.6). In fact the fourth and fifth fish could do better by hunting alone, but it may be that different individuals occupy the lead position during different chases. This is a reminder of the general point that benefits of being in a group may not be shared equally.

HARVESTING RENEWING FOOD

Suppose an animal eats food which renews itself continuously, for example growing vegetation. The amount of food available in a site

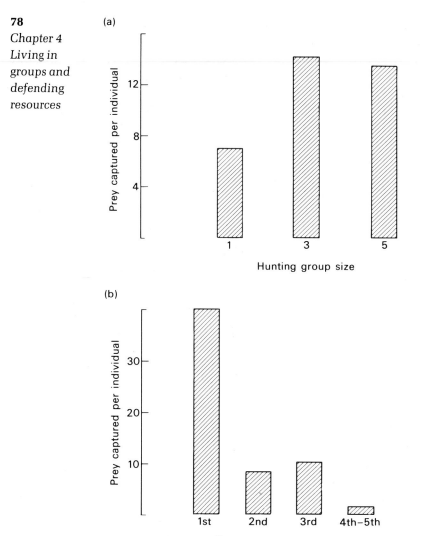

Fig. 4.6 The jack is a predatory fish which hunts in schools. (a) Each fish on average captures more prey per experiment when hunting in a group. (b) But fish at the front of the group benefit the most (Major 1978).

increases with time since the last visit, so an individual could get the maximum possible foraging returns by coming back to the same site after the appropriate time interval. Returning too soon means not finding enough food and returning too late means missed opportunities to eat a plentiful food supply. The problem with harvesting a renewing food supply in this way is that it only works if there is no interference by others with the renewal pattern. Individual A's strategy of returning after ten days would fail if B visited the site after, for example, nine or eight days. One way to prevent interference by others is to defend a territory (Charnov et al. 1976) and another way is to visit sites in a group so that everyone returns at the same time.

Wintering flocks of Brent geese (*Branta bernicla*) feeding on salt marshes in Holland seem to do the latter, where territory defence would not be feasible because the marsh is frequently inundated at high tide. Continuous observation of 40 one hectare plots from dawn to dusk for 24 days during the spring showed that the flocks return to exactly the same site on the marsh at regular 4 day intervals. Not only does this allow the sea plantain (*Plantago maritima*) to recover between visits by the geese, but also the regular cropping pattern actually stimulates the growth of young leaves which are rich in nitrogen. Experiments in which sea plantain was cut with scissors to simulate goose grazing at different time intervals suggested that given the average bite size, the geese may even return after a time interval which maximises the growth of young shoots (Prins, Ydenberg & Drent 1980).

Do individuals at the back of the group fare less well than those at the front, as in the hunting schools of fish described earlier? The answer is not yet known, but it is possible that the overall benefit in terms of food intake is similar for birds in different parts of the flock. The front birds eat most of the vegetation, but the youngest and most nutritious parts of the *Plantago* plants are the base of the leaves close to the ground. These parts are exposed only after the older, taller, parts of the leaves have been cropped off, so it is reasonable to hypothesise that the first birds eat larger mouthfuls while the later birds eat food of higher quality. The overall effect may be that all birds obtain the same quantity of nutrients.

COSTS ASSOCIATED WITH FEEDING

The goose study suggests an important potential cost of feeding in a group: competition for food. Competition may take the form of direct exploitation as in the jack where fish at the front catch the prey and deprive those at the back of the school, or it may arise as a result of interference in which the availability of food to a group member is reduced as a result of the behaviour of nearby companions. This occurs in the redshank (*Tringa totanus*) a shorebird studied by John Goss-Custard on the coastal mudflats of Britain. Redshank feed in tight flocks at night and more loosely scattered or solitarily during the day: this difference seems to be related to interference. In the daytime, redshank feed by sight on small shrimps (*Corophium*) which live with their tails just sticking out of the mud surface while at night, when visual search is impossible, the birds turn to feeding by touch on snails (*Hydrobia*), sweeping their long beaks through the mud. When feeding on shrimps the birds are likely to interfere with one another because the shrimps retreat into the mud and become unavailable as soon as they detect the heavy clump of redshank feet (this was shown in experiments with captive birds). Feeding rate is therefore greater with increasing neighbour distance and the birds tend to space out. At night there is no feeding interference because the birds do not depend on seeing the prey and the snails in any case do not react quickly to disturbance. Feeding rate in these conditions is not related to flock density and the birds crowd into tight groups (Goss-Custard 1976). One interpretation of these results is that there is an advantage to tight clumping in redshank (perhaps as a consequence of seeking cover from predators in the middle of the flock) and that during the night there is little cost to feeding close together. During the day, however, feeding close together causes interference and the birds spread out. The nearest neighbour distance under these conditions appears to reflect a balance between the costs and benefits of group living.

Weighing up costs and benefits—optimal group size

The message of the chapter so far is that there are many different costs and benefits of living in a group, some or all of which might be relevant to a particular species. Pigeons, horses, ostriches and cicadas do not necessarily get together for the same reasons, but they could do so for any or all of the reasons discussed. Our list has been by no means comprehensive; we could have gone on to describe costs of group living such as transmission of diseases, cannibalism, and cuckoldry, or benefits such as protection from the elements, coopera-tive territorial defence, and increased efficiency of locomotion. (Some of these are summarised in Table 4.1). But rather than try to extend the list of possible costs and benefits to its exhaustive and exhausting conclusion, we will return to the more interesting question of how different kinds of cost and benefit can be combined to predict an optimal group size.

Table 4.1 Examples of studies in which possible costs and benefits of group living other than those mentioned in the text have been measured.

	Benefits	
Hypothesis	Test	Reference
1. Warm-blooded animals save energy because of thermal advantage of being close together	Pallid bats (*Antrozous pallidus*) roosting in groups use less energy than solitary roosters	Trune & Slobodchikoff (1976)
2. Small species can overcome competitive superiority of a large species by being in a group	Groups of striped parrotfish (*Scarus croicus*) can feed successfully inside the territories of the competitively superior damsel fish (*Eupomacentrus flavifros*)	Robertson *et al.* (1976)
3. Hydrodynamic advantage for fish swimming in a school. They save energy by positioning themselves to take advantage of vortices created by others in the group	Measurements of distances and angles between individuals show that they are *not* correctly positioned to benefit according to the predictions of the theory	Weihs (1973) Partridge & Pitcher (1979)

Table 4.1 *continued*

Costs

Hypothesis	Test	Reference
4. Increased incidence of disease as a result of close proximity of others	Measure number of ectoparasites in burrows of prairie dogs (*Cynomys* spp). There are more parasites per burrow in larger colonies	Hoogland (1979b)
5. Risk of cuckoldry by neighbours	In colonial nesting red-winged blackbirds (*Agelaius phoenicus*) the mates of vasectomised males laid fertile eggs. They must have been fertilised by males other than their mates	Bray *et al.* (1975)
6. Risk of predation on young by cannibalistic neighbours	In colonies of Belding's ground squirrels (*Spermophilus beldingi*) females with small territories are more likely to lose their young to cannibalistic neighbours than are females with large territories around their burrows	Sherman (1981)

COMPARATIVE STUDIES

A qualitative picture of how costs and benefits interact can be drawn from comparisons between species. For example, some species of shorebirds such as the knot feed in dense flocks while others such as the ringed plover feed in loose flocks or as solitary birds (Fig. 4.7) It is known that living in a flock confers protection on shorebirds against attacks by birds of prey (Page & Whitacre 1975) so why do not all species feed in tight flocks? The species which feed in dense flocks hunt by touch and walk slowly, probing or swinging their beaks through the mud, while solitary and loose flock feeders hunt by sight and move rapidly, picking prey off the surface of the mud or water. Perhaps, as with the redshank, the costs of feeding interference are so big in the latter species that the net benefit for an individual is higher when alone, even though the risk of predation is greater.

TIME BUDGETS

In order to predict more precisely how different costs and benefits combine to determine group size we will return to the approach of chapter 3. Ultimately the costs and benefits influence survival and reproduction, but as with the optimality models described in chapter 3 it is more useful to think in terms of a proximate currency that ultimately relates to fitness.

Ron Pulliam (1976) and Tom Caraco (1979a) have used time as a currency and developed a model of optimal group size based on time

Fig. 4.7 Knot (*Calidris canutus*) (top) feed by touch and live in dense flocks while ringed plovers (*Charadrius hiaticula*) (bottom) feed by sight and live in loose locks or as solitary individuals (Goss-Custard 1970). The difference can be interpreted in terms of the costs and benefits of flocking.

budgets. The model is meant to illustrate the factors influencing winter flocks of small birds. The survival of birds in a flock is considered to be dependent on two main risks, starvation and predation, and the birds' time budget is divided into three types of behaviour associated with these risks: scanning (for predators), feeding, and fighting (for food). Based on their observations of yellow-eyed junco flocks, Pulliam and Caraco divide fighting into two categories: short term squabbles over access to pieces of food and attacks in which dominant birds attempt to evict subordinates from good feeding sites in order to ensure a supply of food for the rest of the winter. The three activities in the time budget are assumed to be mutually exclusive, that is a bird cannot, for example, scan and feed at the same time. In order to scan it has to point its head upwards, while pecking involves facing towards the ground. Finally they assume that scanning for predators takes precedence over feeding, since failing to see an approaching predator is more dangerous than failing to eat a seed. Dominant birds are assumed to give higher priority to satisfying their daily energy requirements than to long term eviction of subordinates,

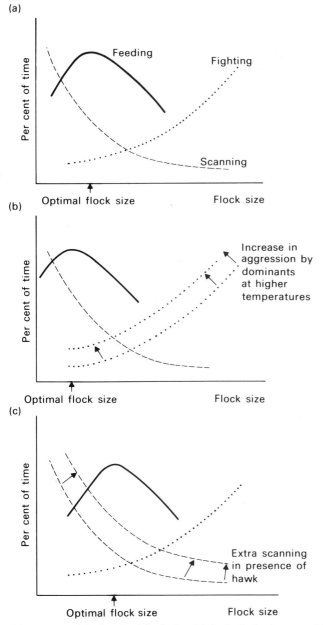

Fig. 4.8 A model of optimal flock size. (a) As flock size increases birds spend more time fighting and less time scanning. An intermediate flock size gives the maximum proportion of time feeding. (b) At higher temperatures (or when food is more plentiful) dominant birds can afford to spend more time attacking subordinates. The optimal flock size for the average bird therefore decreases. (c) When predation risk is increased by flying a hawk over the flock, the scanning level should go up and the optimal flock size is increased (based on Pulliam 1976; Caraco *et al.* 1980b).

while for a subordinate bird aggression must take priority over feeding since a bird cannot feed while it is being attacked. Figure 4.8a shows a simplified version of Pulliam's and Caraco's model. The main features are as follows.

1   The proportion of time spent scanning by an individual is assumed to decrease with increasing group size. The basis for this assumption is that a given level of vigilance can be maintained with less scanning per individual as group size increases (p. 71).

2   As group size increases and encounters between birds become more frequent, the proportion of time spent in aggression increases.

3   The time spent feeding therefore is at a maximum in flocks of intermediate size.

Can this model of time budgets be used to predict the optimal group size? If the only benefit of flock feeding is to increase the time available for feeding while maintaining a certain level of vigilance, the optimal flock size is the one indicated in Fig. 4.8a. If there are other benefits of flocking such as dilution and increased vigilance (see pp. 71–2), the optimal flock size may be larger than the one shown in Fig. 4.8a. Therefore the model can be used to test whether or not maximising food intake is the only benefit of flocking. However the picture is probably more complicated than Fig. 4.8a suggests, because the optimal flock size may be different for dominant and subordinate birds. Dominant birds obtain a long term benefit from evicting subordinates, so they should prefer to be in smaller groups.

Caraco (1979b) and Caraco *et al.* (1980) have tested some of the assumptions of their model by recording the time budgets of yellow-eyed juncos in winter flocks in Arizona. They found that the proportion of time spent by individuals in scanning and fighting change with flock size in the directions assumed by the model. However the decrease in scanning time was much greater than the increase in fighting time over the range of flock sizes they studied, so that feeding time increased with flock size, as it does to the left of the peak in Fig. 4.8a.

In order to test whether time budgets influence flock size in the way suggested by the model Caraco *et al.* predicted the effect of various environmental changes on flock size. The predictions were as follows.

1   As average daily temperature increases dominant birds should have more time to evict subordinates because they can satisfy their energy requirements more rapidly. Flock size should therefore decrease (Fig. 4.8b). This prediction was supported by observations: at 2°C flocks contained an average of 7 birds, at 10°C they contained 2 birds. This decrease coincided with an increase in time spent fighting by dominant birds.

2   By a similar argument, an increase in food supply should produce a decrease in flock size and an increase in the proportion of time spent fighting by dominant birds. Again the results of field observations supported the prediction. When food was scattered in the canyon, the birds fed in smaller flocks.

3   An increase in the risk of predation should have exactly the opposite effect of the previous two changes. This is because a high risk of predator attack should cause the birds to spend more time scanning;

they therefore have to feed in larger flocks to maintain a given rate of food intake (Fig. 4.8c). Caraco *et al.* (1980) allowed a tame hawk to fly over the canyon and as predicted the birds spent more time scanning and the mean flock size increased. It was 3.9 birds without the hawk and 7.3 with the hawk.

4 Finally, Caraco predicted that by adding more cover to the canyon in the form of a bush, the effective risk of predator attack would be reduced because the juncos would have easier access to a safe hiding spot. The birds should therefore spend less time scanning, which allows more time for feeding and fighting. This is exactly what happened when an experimental bush was placed near one of the favoured feeding sites, and as expected the flock size decreased.

What can we conclude from these results? First they show that flock size is influenced by time budgets as hypothesised in the model shown in Fig. 4.8a. Flocking allows more time for feeding because less time is spent scanning, and the maximum flock size depends on the time available for dominant birds to evict subordinates. Second, the results allow us to reject the simplest hypothesis about optimal flock size. The birds did not feed in flocks of the size that would have maximised feeding time: under normal conditions the average flock in the canyon contained 3.9 birds, but measurements showed that the time available for feeding would have been higher in a flock of 6 or 7. As we have mentioned already, the optimal flock size for dominant and subordinate birds probably differs since dominant birds benefit by evicting subordinates. The observed flocks may have been a compromise between the optimum for dominant and subordinate birds. A further complication is that birds in larger flocks benefit by the dilution effect and increased vigilance, as described earlier for ostriches (Caraco *et al.* 1980).

The model in Fig. 4.8 is clearly too simple, but the study shows that time budgets can be used to analyse the effects of different costs and benefits on flock size. It also introduces the idea that flocking and resource defence may be two ends of a continuum. The model could be viewed as one which predicts the conditions under which it pays dominant birds to exclude subordinates and defend a territory. When food is plentiful or predation risk is low dominants can afford the time to maintain a defended area, or in other words the territory is economically defendable.

Resource defence

Many animals ranging from sea anenomes to monkeys defend resources such as mating sites, feeding areas, and nests against competitors. Defence of a resource by fighting or displays is usually described as territorial behaviour. Why should animals compete for resources by defence rather than by direct exploitation? The first attempts to answer this question consisted of drawing up lists of the 'functions' of territories (Hinde 1956; Tinbergen 1957). This approach was valuable in emphasising the diversity of kinds of territory—some animals such as sheet web spiders defend small, temporary areas for mating, others such as European blackbirds defend large areas for a whole lifetime, and so on—but it did not lead to a general answer to the question of why animals defend territories.

ECONOMIC DEFENDABILITY

Jerram Brown (1964) first introduced the idea of economic defendability. He pointed out that defence of a resource has costs (energy expenditure, risk of injury and so on) as well as the benefits of priority of access to the resource. Territorial behaviour should be favoured by selection whenever the benefits are greater than the costs (Fig. 4.9). This may seem a rather obvious conclusion and indeed as stated so far it is. However it led field workers to look in more detail at the time

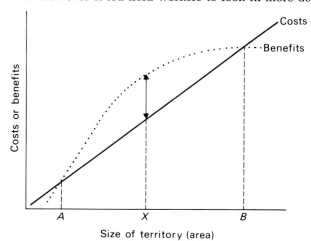

Fig. 4.9 The idea of economic defendability. As the amount of resource defended (or territory size) increases so do the costs of defence. The benefits (e.g. the amount of food available) will increase at first but level off as the resource becomes superabundant for the animal. The resource is economically defendable between A and B. Within this range maximum net gain (for example of food) is at X and the minimal cost is at A.

budgets of territorial animals, in particular the feeding territories of nectivorous birds. While Brown's idea is in principle applicable to any kind of territory, it has been most useful in looking at birds such as hummingbirds, sunbirds, and honeycreepers in which the costs and benefits can be measured in calories in the field. Frank Gill and Larry Wolf (1975), for example, were able to measure the nectar content of territories of the golden-winged sunbird (*Nectarinia reichenowi*) in East Africa, where it defends patches of *Leonotis* flowers outside the breeding season. They also calculated from time budget studies and laboratory measurements of the energetic cost of different activities such as flight, sitting, and fighting how much energy a sunbird expends in a day. When the daily costs were compared with the extra nectar gained by defending a territory and excluding competitors, it turned out that the territorial birds were making slight net energetic profit. The resource was economically defendable (Box 4.1).

Box 4.1 *The economics of territory defence in the golden-winged sunbird. (Gill & Wolf 1975).*

(a) The metabolic cost of various activities was measured in the lab:

Foraging for nectar 1000 cal/hour
Sitting on a perch   400 cal/hour
Territory defence   3000 cal/hour

(b) Field studies showed that territorial birds need to spend less time per day collecting enough energy in the form of nectar to survive when the flowers contain more nectar:

Nectar per flower (µl)	Time to get energy (h)
1	8
2	4
3	2.7

(c) By defending a territory a bird excludes other nectar consumers and therefore increases the amount of nectar available in each flower. The bird therefore saves foraging time because it can satisfy its energy demands more rapidly. It spends the spare time sitting on a perch, which uses less energy than foraging. For example, if defence results in an increase in the nectar level from 2µl to 3µl per flower, the bird saves 1.3 h per day foraging time [from (b)]. It therefore saves:

$(1000 \times 1.3) - (400 \times 1.3) = 780$ calories
  foraging   resting

(d) But this saving has to be weighed against the cost of defence. Measurements in the field show that the birds spend about 0.28 hours per day in defence. This time could otherwise be spent sitting, so the extra cost of defence is:

$(3000 \times 0.28) - (400 \times 0.28) = 728$ calories

In other words the flowers are just economically defendable when the nectar levels are raised from 2 to 3µl as a result of defence. Gill and Wolf found that most of their sunbirds were territorial when the flowers were economically defendable.

The idea of economic defendability has also been used to predict the levels of resource availability which could lead to territorial defence. If resources are very scarce, the gains from excluding others may not be sufficient to pay for the cost of territorial defence. Instead the animal might abandon its territory and move elsewhere. There may also be an upper threshold of resource availability beyond which defence is not economical. This upper boundary could arise for a number of reasons.

1 There may be so many intruders trying to invade rich areas that defence costs would be prohibitively high (Fig. 4.1b).

2 There may be no advantage of territoriality at high resource levels if the owner cannot make use of the additional resources made available by defence. In Gill and Wolf's sunbirds, one advantage of territorial defence was that it raised the amount of nectar per flower (by exclusion of nectar thieves) and hence saved foraging time (Box 4.1). But if nectar levels are already high, the extra increment resulting from territorial defence saves hardly any foraging time. This is because the bird's rate of food intake at high nectar levels is limited by the time it takes to probe its beak into the flower (the handling time). For example Gill and Wolf calculate that an increase from 4μl to 6μl per flower would save the birds less than 0.5 h of foraging time, while as shown in Box 4.1 an increase of from 1μl to 2μl saves 4 h. Thus when nectar levels are high, territorial exclusion of nectar thieves does not pay for itself in savings of foraging time.

3 A third hypothesis which predicts an upper boundary was proposed by Carpenter and McMillen (1976). They suggested that territory defence has associated risks such as increased conspicuousness to predators, so that whenever resource levels are high enough to allow an animal to satisfy its needs without excluding others, territories should be abandoned. As yet there is no critical evidence to test this idea.

OPTIMAL TERRITORY SIZE

Although the use of Brown's concept to predict the range of resource levels over which animals should defend territories is a step forward from simply showing that territories are economically defendable, a more quantitative and powerful development is to predict the optimal amount of resource for an individual to defend. Gill and Wolf found that their sunbirds always defended about 1600 flowers, even though the territory area varied by 300 fold! Graham Pyke (1979b) has tried to find out why sunbirds defend 1600 flowers by building an economic model of optimal territory size. He used Gill & Wolf's data to calculate the time budgets and energy expenditures of birds defending different numbers of flowers, assuming that a territory with more flowers and therefore more nectar would have more intruders. As we have seen before, an optimality model involves choosing a criterion of benefit which is to be maximised. Pyke considered four possibilities. The birds might choose a territory size which maximises daily net energy gain (X in Fig. 4.9): this could be a good measure of benefit for a bird that is short of food or one that is storing food for future hard times. The second idea was that the birds might maximise the _ratio_ of gains to costs. A third hypothesis was that the birds might maximise the daily sitting time, a reasonable goal if a sitting bird is much safer from predators than one flying around in the open. The fourth hypothesis was that the birds would minimise their daily energy costs (A in Fig. 4.9). The rationale for this is that metabolising energy wears out the machinery of the body, so that an animal intent on lasting out until the next breeding season should minimise costs. Pyke's calculations showed that the third hypothesis accurately predicts both the number of flowers defended and the daily time budget of the sunbirds (Box 4.2). The second hypothesis is good at predicting the number of flowers defended but not the time budget.

Box 4.2 Pyke's model of optimal territory size in sunbirds.

1 The bird's time budget consists of four components:

$$\text{sleeping} + \text{feeding} + \text{sitting} + \text{defence}$$

$$Z \qquad F \qquad S \qquad D \qquad \text{(i)}$$

2 The total daily energy costs (C) are made up of the 'fixed cost' of sleeping (Z) and the variable costs of F, S and D. These are considered to be variable because the bird can alter its time budget to change the total cost.

3 The total daily gain (G) from the territory is:

$$\text{standing crop of energy/flower} \times \text{rate of visiting}$$

$$\times \text{ feeding time/day} = erF \qquad \text{(ii)}$$

Pyke assumes that the food supply is in equilibrium, i.e. the birds eat as much as is produced each day. The daily production is given by

number of flowers $\times$ daily production/flower

$= np$

$G = erF = np$

$$e = \frac{np}{rF}. \qquad \text{(iii)}$$

4 The defence time is assumed to be related to the quality of the territory. More invaders try to get into better quality territories.

$$D = kne^{\alpha} = \frac{k'n^{\alpha+1}}{F^{\alpha}} \text{ (by substitution of iii)} \qquad \text{(iv)}$$

(k and k' are constants, α determines the form of the relationship between D and the standing crop (ne). If $\alpha > 1$, the curve of D against ne is accelerating).

5 Pyke's hypotheses are that the birds might choose n and F (size of territory, measured as number of flowers defended, and feeding time) so as to maximise $E = G - C$, or minimise C, or maximise S, or maximise G/C. All of these have to be subject to the constraint that $E \geqslant 0$, otherwise the bird would die of starvation! Note that the choice of *n* and *F* completely determines the time budget: *n* and *F* determine *D* (equation iv) and hence *S* (by equation i). This in turn determines the value of *C* (see point 2 above).

6 Pyke used laboratory measurements of the costs per minute of *F,S,D* and *Z* and field estimates of *e* and *n*. He assumed that the value of α lies between 1 and 3 and calculated *k'* from equation (iv).

7 With α set equal to 2 the various hypotheses made the following predictions:

Four different hypotheses

	Max E	Max G/C	Max S	Min C	Observed
Number of flowers	7070	6722	1653	1595	1600
Foraging time (h)	7.45	8.18	1.72	2.41	2.42
Defence time (h)	2.55	1.82	0.61	0.28	0.28
Sitting time (h)	0	0	7.67	7.31	7.30

Both maximising *S* and minimising *C* predict approximately the correct number of flowers but the latter is better at predicting the time budget. The reason is that defence costs are very high relative to other activities, therefore minimising *C* produces a low value of defence time. However maximising *S* takes no account of the cost of defence, so it predicts a higher value of *D*.

One way to test Pyke's model further would be to measure the relationship between territory quality and defence time. Pyke predicts that α should be about 2.

This example shows how the concept of economic defendability can be developed into a quantitative argument about optimal territory size. Up to now this approach has been used only for feeding territories, but it will be an interesting development to apply it to other kinds of territory. The major limitation of this approach is one mentioned at the end of chapter 3. Often the best strategy for an individual depends on the strategy adopted by the rest of the population. The optimal amount of resource to defend may vary according to the amount defended by others. In order to analyse this question we have to pose the problem as follows: 'if all individuals in the population defend a territory of size *X*, could a single "mutant" individual defending a bigger or smaller territory than *X* do better?' In the next chapter we will develop this approach.

Summary

91

*Chapter 4
Living in
groups and
defending
resources*

Living in groups may help to protect the individual against predators and enhance its ability to find or capture food. There are also costs of group living such as increased conspicuousness and competition for food. The size of group in which animals live should reflect a balance between advantages and disadvantages, but there may be a different balance for different individuals within a group. One way of combining various kinds of cost and benefit is to use the common currency of time and predict optimal group sizes on the basis of time budgets.

There is often a continuum between group living and resource defence, with individuals becoming territorial when resources are economically defendable. The idea of economic defendability can be used to study the time budgets of territorial animals and predict how much resource they should defend.

Further reading

Living in groups:

Bertram (1978) and Wilson (1975 chapter 3) are good reviews of why animals live in groups.

Hoogland (1979a, b) is a study of prairie dogs in which both the costs and the benefits of group living are investigated.

Morse (1970) describes some of the reasons why birds sometimes live in mixed species groups and Rubenstein (1978) discusses in general terms how advantages of grouping with respect to food and predators may interact.

Kruuk (1972) is a beautiful field study of the spotted hyena, a group living predator. Barnard (1980) describes the advantages of flocking in sparrows: birds in a flock spend less time scanning and more time feeding.

Defending resources:

Myers *et al.* (1980) provide a detailed review of the economics of resource defence and optimal territory size.

Verner (1977) develops the idea that animals might defend more resources than they actually need to maximise their own success, simply in order to reduce the success of others (superterritories). Appealing though it is, Verner's idea has been criticised on theoretical grounds by several authors (e.g. Rothstein 1978; Parker & Knowlton 1980). The essence of the criticism is that defending a superterritory might pay when the trait is rare, but it will not pay when most of the population adopts this strategy.

Davies and Houston (1981) analyse the costs and benefits of territory sharing in the pied wagtail (*Motacilla alba*) and show how as food abundance increases, a territory owner does better by sharing its territory with a satellite.

Chapter 5. Fighting and Assessment

So far in this book we have emphasised the role of ecological factors in determining an animal's behaviour. In particular, we have seen the importance of food and predators. However, there is another influence on an individual that we have ignored, namely the behaviour of other competitors in the population.

When an animal decides where to search for a scarce resource such as food, a mate or a territory, its decision must depend on what all the other individuals in the population are doing. When an animal competes directly with others, its decision of whether to fight or retreat will depend on the behaviour and strength of its opponents. In this chapter we will examine how the behaviour of other competitors determines the evolution of an individual's searching and fighting strategies.

The problem of where to search

Consider two extreme models for resource competition (Fig. 5.1). In case A there is a limit to the number of competitors who can gain access to the resource so that some individuals will be excluded by the despotic behaviour of others. In case B there is resource sharing so that all individuals get access to the resource.

CASE A: DESPOTISM

Imagine that there are two types of habitat, one rich in resources and one poor. The first competitors to arrive will go to the rich habitat and then, once this is filled, others will be forced to occupy the poor habitat. When this is also full, any further individuals will be excluded from the resource altogether (Fig. 5.2).

This kind of situation is very common in nature. In Wytham Woods (see p. 13) the best breeding habitat for great tits is in oak woodland. This is quickly occupied in the spring and becomes completely filled with territories. Some individuals are excluded from the oak wood and have to occupy the hedgerows nearby where there is less food and consequently lower breeding success. If great tits are removed from the best habitat then birds rapidly move in from the hedgerows to fill the vacancies (Krebs 1971). Similarly, in red grouse (*Lagopus lagopus scoticus*) territorial birds defend the richest areas of the heather moors as breeding and feeding territories. Excluded birds have to go about in flocks and exploit poor habitats where their chances of survival are

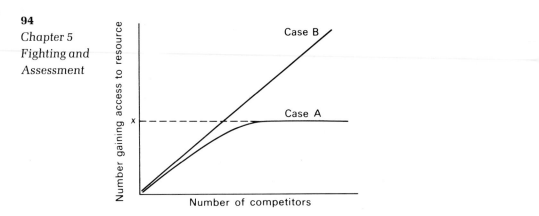

Fig. 5.1 Two simple models of resource competition. In case A some individuals are despots and they exclude other competitors from the resource. Only x individuals are able to gain access to the resource. In case B there is resource sharing so that all competitors enjoy access to the resource.

low. Once again, if a territory owner is removed its place is quickly taken by a bird from the flock (Watson 1967).

In these examples the strongest individuals are despots, grabbing the best quality resources and forcing others into low quality areas or excluding them from the resource altogether.

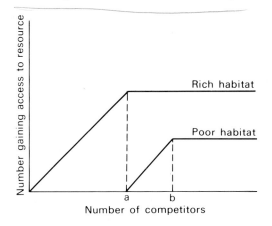

Fig. 5.2 Case A, despotic behaviour. Competitors occupy the rich habitat first of all. At point a this becomes full and newcomers are now forced to occupy the poor habitat. When this is also full (point b), further competitors are excluded from the resource altogether and become 'floaters' (after Brown 1969).

CASE B: RESOURCE SHARING

Again imagine two habitats, one rich and one poor, but this time there is no despotic behaviour. All individuals are free to go wherever they like and enjoy access to the resource. This may sound like an ideal world, and is referred to as 'ideal free conditions' (Fretwell & Lucas 1970).

In this case the first competitors to arrive will go to the rich habitat. There is no territoriality or fighting and so there is no limit to the number of individuals who can go there. However the more competitors that occupy the rich habitat the more the resource will be depleted, and so the less profitable it will be for further newcomers. Eventually a point will be reached where the next arrivals will do better by occupying the poorer quality habitat where, although the resource is in shorter supply, there will be less competition (Fig. 5.3). Thereafter the two habitats will be filled so that the profitability for an individual is the same in each one. In an ideal free distribution competitors adjust their distribution in relation to habitat quality so that each individual enjoys the same gain.

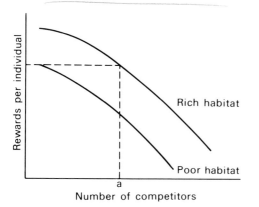

Fig. 5.3 Case B, resource sharing. There is no limit to the number of competitors who can exploit the resource. Every individual is free to choose where to go. The first arrivals will go to the rich habitat. Because of resource depletion, the more competitors the lower the rewards per individual so at point *a* the poor habitat will be equally attractive. Thereafter the habitats should be filled so that the rewards per individual are the same in both (after Fretwell 1972).

We can see a good example of this in action when people stand in line at the counters of a supermarket. If all the serving clerks are equally efficient and all the customers are equal in the time they require for service, then the lengths of all the lines should end up being equal. If one line gets shorter, then customers would profit by joining it until its length becomes the same as the others. Because everyone is free to join whichever line they like, each person goes to the best place at the time and the lines fill up in an ideal free way with the result that every customer should have the same waiting time for service.

An exact equivalent in animals is Manfred Milinski's (1979) experiment with sticklebacks. Six fish were put in a tank and prey (*Daphnia*) were dropped into the water from a pipette at either end. At one end prey were dropped into the tank at twice the rate of the other end. The best place for one fish to go depends on where all the others go. There was no resource defence (despotism) and Milinski found that the fish distributed themselves in the ratio of the patch profitabilities, with four fish at the fast-rate end and two at the slow-rate end. When the feeding regimes were reversed, the fish quickly redistributed themselves so that four were again at the fast end (Fig. 5.4). This is the only stable distribution under ideal free conditions. With any other distribution it would pay an individual to move. For example, if there were three fish at each end then one fish would profit by moving from the slow to the fast-rate end. Once it had done so, it would not pay any of the other fish to move. In our supermarket analogy, this experiment is equivalent to what should happen if one clerk is twice as efficient at serving customers as another; the stable distribution would be for this line to be twice as long.

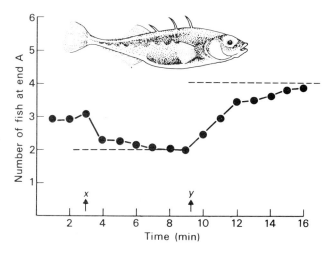

Fig. 5.4 Milinski's (1979) feeding experiment with six sticklebacks. At point x, end B of the tank had twice the amount of food as end A. At point y the profitabilities were reversed. The dashed lines indicate the number of fish predicted at end A according to ideal free theory, and the solid line is the observed numbers (mean of several experiments).

The stable distribution of searching individuals could be achieved in two ways. For example, if one habitat was twice as profitable as another then stability could come about by:

1 Competitor numbers adjusting so that twice as many individuals go to the good habitat as to the poor one.

2 All individuals visiting both habitats, but each spending twice as much time in the good habitat as the poor one.

IDEAL FREE DISTRIBUTIONS AND DESPOTISM

Most examples in nature will have features of both the simple models we have discussed above. Perhaps the commonest situation will be where the best place to search depends on where all the other competitors are but within a habitat some individuals get more of the resource than others. In the fish experiment, for example, our population counts may show a stable, ideal free distribution of individuals but the chances are that some fish will be better competitors than others. At each end of the tank there could be one or two large fish grabbing most of the prey. The ideal free distribution could come about because of the way the subordinates distribute themselves in relation to the despots. In effect, the despots are part of the habitat to which the subordinates respond when deciding where to search.

It is unlikely that there will be any populations where all individuals are of equal competitive ability. Even though male dungflies obey an ideal free distribution around cow pats, the larger males get more females than small males (Borgia 1979). In our supermarket there is probably no fighting between the customers but some will have more items of shopping than others and will require more time to be served. In a sense, they are greater competitors because standing in line behind them will impose greater waiting times.

A good example which shows features of both the despotic and ideal free models is the study by Thomas Whitham (1978, 1979, 1980) of habitat selection in the aphid (*Pemphigus betae*). In the spring females, known as 'stem mothers', settle on leaves of narrowleaf cottonwood (*Populus angustifolia*) to feed and they become entombed by expanding leaf tissue, so forming a gall. A stem mother reproduces parthenogenetically and the number of progeny she produces depends on the quantity and quality of the juices she can tap from the leaf. The largest leaves provide the richest supplies of vascular sap and result in the greatest reproductive success, with up to seven times the number of progeny that are produced by settling on a small leaf. As we would expect, all the large leaves are quickly occupied and so additional settlers have the problem of whether to settle on large leaves and share the resources or occupy smaller leaves alone.

Whitham made measurements of reproductive success which enabled him to plot a family of fitness curves for habitats of varying quality (leaves of different sizes) and with different densities of competitors (number of galls per leaf). Figure 5.5 shows the results, which enables us to draw three conclusions. First, for any competitor density, the average reproductive success increases with habitat quality. Second, within a habitat of a certain quality, success decreases as the number of competitors increases. This shows that stem mothers settling on the same leaf must compete with each other for resources. Third, if the *average* reproductive success is calculated for aphids who are alone on a leaf, those who share a leaf with one other and those who share with two others, no significant differences are found. There was also no significant difference in average success on leaves with different numbers of competitors when other fitness measures were used such as body weight of the stem mother, abortion rate, development rate or predation. The results support the predictions of the ideal free model. The conclusion, therefore, is that the stem mothers settle

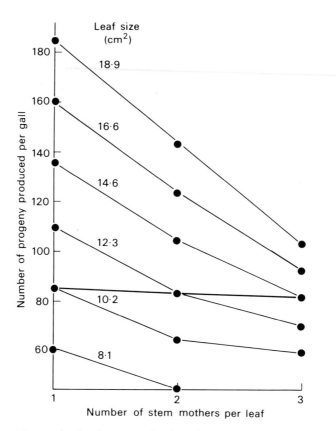

Fig. 5.5 The thin lines are a family of fitness curves for habitats of varying quality (leaf-size) and competitor density (no. stem mothers per leaf) in the aphid *Pemphigus betae*. The solid horizonal line is the average success for one, two and three stem mothers per leaf. See text for explanation (from Whitham 1980).

on leaves of different sizes such that the average success in good habitats with a high density of competitors is the same as in poor habitats with fewer competitors.

However, although the results of *average* success on different sized leaves are in accord with ideal free predictions, within a habitat not all individuals get equal rewards. This is because a leaf is not a homogeneous habitat. The best place to be is on the midrib at the base of the leaf blade because everything translocated into and out of the leaf must flow past this point. Basal galls on a leaf give rise to more young than distal galls and the stem mothers spar with each other, like boxers in a ring, for occupancy of these prime positions (Fig. 5.6). As we would predict from the despotic model, if a basal individual is removed her place is quickly occupied by another aphid from a distal site.

In the aphid contests the pushing matches appear to be trials of strength and it is usually the larger stem mother that wins the fight. In general, because not all individuals within a population will be of equal competitive ability, we would expect an animal to assess

Fig. 5.6 Stem mother aphids *Pemphigus betae* fight for prime positions on a leaf by kicking and pushing. The winner will settle at the base of the mid-rib where food is richest (Whitham 1979).

competitor strength as well as competitor numbers when deciding which habitat to occupy. How should an individual behave when it comes into direct conflict with another over some scarce resource? Should it fight or retreat? In the next section we shall see that the evolution of fighting strategies will depend on what others in the population are doing.

The problem of how to fight

When animals compete for an item of food, a territory or a mate, they often display to one another; stags roar, frogs croak, fish wave their fins and birds fluff out their feathers. Sometimes the competitors come to blows but usually, after a brief display, one retreats and the other gains possession of the resource without a serious fight. How do individuals decide who will win and who will loose? Why do animals often settle disputes through conventional displays rather than by all out fighting?

The long-accepted answer to this question was that escalated contests would result in many animals getting seriously injured and this would militate against the survival of the species (e.g. Lorenz 1966; Huxley 1966). This is obviously a group selection argument and does not explain how natural selection acting on individuals can give rise to the evolution of conventional fighting. Since the mid 1960's many people realised that the answer must be in terms of the costs and benefits of fights to individual animals. However it is only during the last few years that Maynard Smith and others have formalised how contest behaviour can evolve by the selection of individual strategies (Maynard Smith & Price 1973; Maynard Smith 1976b, 1979; Maynard Smith & Parker 1976). The main thrust of the new models is that it does not make sense to ask what is the best way for an individual to fight, without considering how other individuals in the population are behaving.

Imagine that evolution proceeds rather like a game, with individuals playing various strategies. A strategy is defined as a pre-programmed behavioural policy, one of a set of possible alternatives. As a simple hypothetical example, consider a game where there are just two sorts of strategies. 'Hawks' always fight to injure and kill their opponents, though in the process they will risk injury themselves. 'Doves' simply display and never engage in serious fights. These two strategies are chosen to represent the two possible extremes that we may see in nature.

In this evolutionary game, let the winner of a contest score +50, and the loser 0. The cost of a serious injury is −100 and the cost of wasting time in a display is −10. These pay-offs are some measure of fitness and we will assume, for simplicity, that Hawk and Dove reproduce their own kind faithfully in proportion to their pay-offs. (The exact values do not matter and are chosen simply because this game is easier to explain with numbers rather than algebra.) The next step is to draw up a two by two matrix with the average pay-offs for the four possible types of encounter. These calculations are explained in Table 5.1.

Table 5.1 The game between Hawk and Dove (after Maynard Smith 1976b).
(a) Pay-offs: Winner +50   Injury −100
                     Loser   0   Display − 10
(b) Pay-off Matrix: average pay-offs in a fight to the attacker.

Attacker	Opponent	
	Hawk	Dove
Hawk	(a) $\frac{1}{2}(50) + \frac{1}{2}(-100)$ $= -25$	(b) +50
Dove	(c) 0	(d) $\frac{1}{2}(50 - 10) + \frac{1}{2}(-10)$ $= +15$

Notes
(a) When a Hawk meets a Hawk we assume that on half of the occasions it wins and on half the occasions it suffers injury.
(b) Hawks always beat Doves.
(c) Doves always immediately retreat against Hawks.
(d) When a Dove meets a Dove we assume that there is always a display and it wins on half of the occasions.

How would evolution proceed in this particular game? Consider what would happen if all individuals in the population are Doves. Every contest is between a Dove and another Dove and the pay-off is on average +15. In this population, any mutant Hawk would do very well and the Hawk strategy would soon spread because when a Hawk meets a Dove it gets +50. It is clear that Dove is not an evolutionarily stable strategy, or ESS.

However, Hawk would not spread to takeover the entire population. In a population of all Hawks the average pay-off is −25 and any mutant Dove would do better because when a Dove meets a Hawk it gets 0 (which is not very good, but still better than −25!). The Dove strategy would spread if the population consisted mainly of Hawks. Therefore Hawk is not an ESS either.

Nevertheless, a mixture of Hawks and Doves could be stable. The stable equilibrium will be when the average pay-offs for a Hawk are equal to the average pay-offs for a Dove. If the population moved away from this equilibrium then either Dove or Hawk would be doing better and so the population would not be stable. Each strategy does best when it is relatively rare and the tendency in this evolutionary game will be for frequency dependent selection to drive the frequencies of Hawk and Dove in the population so that they each enjoy the same success. For the values in Table 5.1, the stable mixture can be calculated as follows.

Let h be the proportion of Hawks in the population. Therefore the proportion of Doves must be $(1 - h)$. The average pay-off for a Hawk is the pay-off for each type of fight multiplied by the probability of meeting each type of contestant.
Therefore,

$\bar{H} = -25h + 50(1 - h)$.

Similarly, for Dove the average payoff will be,

$\bar{D} = 0h + 15(1 - h)$.

At the stable equilibrium (the ESS), $\bar{H}$ is equal to $\bar{D}$. Solving the two equations above by setting $\bar{H} = \bar{D}$ gives $h = 7/12$, and therefore, by subtraction, the proportion of Doves $(1 - h)$ must be $5/12$.

The ESS could be achieved in two distinct ways:

1 The population could consist of individuals who played pure strategies. Each individual would either be Hawk or Dove and the ESS would come about with $7/12$ of the population being Hawks and $5/12$ Doves.

2 The population could consist of individuals who all adopted a mixed strategy, playing Hawk with probability $7/12$ and Dove with probability $5/12$, choosing at random which strategy to play in each contest. Both these would produce stability in the game. If the population consisted of a different mixture of individuals from that in (**1**) or of individuals playing their strategies with different probabilities from those in (**2**), then there would be no equilibrium; either Hawk or Dove would enjoy a temporary increase in success until the population evolved to the ESS once more.

HAWK, DOVE AND BOURGEOIS

Now imagine another strategy in this game, 'Bourgeois'. With this strategy the individual plays 'Hawk, if owner' and 'Dove, if intruder'. In other words, it fights hard if it's owner but always retreats if it's the intruder. Let us keep the same pay-offs as before and, for simplicity, imagine that a 'Bourgeois' individual finds itself owner half the time and intruder half the time. With three strategies in the game, the pay-offs are indicated in Table 5.2.

Table 5.2 The Hawk, Dove, Bourgeois game (after Maynard Smith 1976b).
(a) Pay-offs (as in Table 5.1)

Winner +50   Injury −100
Loser   0   Display − 10

(b) Pay-off Matrix: average pay-offs in a fight to the attacker.

Attacker		Opponent	
	Hawk	Dove	Bourgeois
Hawk	−25	+50	+12.5
Dove	0	+15	+7.5
Bourgeois	−12.5	+32.5	+25

Notes
1   The top left four cells are exactly the same as in Table 5.1.
2   When Bourgeois meets either Hawk or Dove we assume it is owner half the time and therefore plays Hawk, and intruder half the time and therefore plays Dove. Its pay-offs are therefore the average of the two cells above it in the matrix.
3   When Bourgeois meets Bourgeois on half the occasions it is owner and wins while on half the occasions it is intruder and retreats. There is never any cost of display or injury.

    In this game, Bourgeois is an ESS. If all the population are playing this strategy, no-one ever engages in escalated fights because when two individuals contest for a resource, one is owner and the other is intruder; the result is that the intruder always gives way. With everyone playing the Bourgeois strategy the average pay-off for a contest is +25. This is stable against invasion by Hawks, who would only get +12.5, and also stable against Doves, who would only get +7.5. In fact Bourgeois is the only ESS in the game in Table 5.2. If all the population played Hawk, both Dove and Bourgeois could invade and do better. If all the population played Dove, then both Hawk and Bourgeois could invade.

SIMPLE MODELS AND REALITY

These simple models are so far removed from what actually happens in nature that we may well ask what use they can possibly be in helping us to understand real animal contests. There are three main conclusions from Maynard Smith's approach.
1   The most important point is that the best fighting strategy for any one individual must depend on what other competitors are doing, because the pay-offs for employing a strategy will be frequency dependent. Is Hawk a good strategy? The answer is yes if the population consists mainly of Doves and no if the population consists mainly of

Hawks. Instead of asking whether a strategy is a *good* strategy, we should really instead be asking is it a *stable strategy, or ESS?*

2 The ESS will depend on the strategies in the game. In the first example, where there were just two strategies, the ESS was a mixture of Hawk and Dove. However, when we introduced another strategy, Bourgeois, into the game the ESS solution changed; Bourgeois turned out to be a pure ESS.

These three strategies are undoubtedly too simple to represent in detail the strategies that animals adopt in the wild. Nevertheless they are plausible alternatives which we could regard as simplified versions of strategies we might see in nature. It is interesting to find that neither Hawk nor Dove is an ESS in our simple games, but some mixture of Hawk and Dove behaviour can be stable. This is exactly what we see in real animals, a mixture of display and fighting. We are not suggesting that the models mimic perfectly evolution in the wild, only that they are a useful tool for gaining some insights into how contest behaviour might evolve.

More detailed models will have to consider a greater variety of more complicated strategies and we will have to rely on biological intuition to help us define the range of possibilities. ESS models cannot tell us what strategies will evolve, only what will be stable given a defined set of alternatives. There is no doubt that an animal with a machine gun would invade the Hawk–Dove game and soon spread, but until we see real animals behaving like this there is little point in putting the strategy into our models.

3 The ESS will also depend on the values of the pay-offs in the game. If we changed the pay-offs in Table 5.1 then the ESS mixture of Hawks and Doves would change. In fact, in this game it can be shown that as long as the cost of injury exceeds the value of winning then neither pure Hawk nor pure Dove can be an ESS. Nevertheless the relative amount of Hawk-like behaviour we would expect must depend on the costs and benefits in the contest.

Therefore if game theory models are going to make precise predictions about contest behaviour, we will not only need to know the range of possible strategies but also the costs and benefits for the pay-off matrix. In the real world, ecological circumstances such as resource abundance and competitor density will determine the pay-offs in the evolutionary game. In practice it is not going to be easy for the field worker to go out and measure on a common scale (effect on fitness) the cost of displaying, the cost of serious injury and the value of winning. However we can see in an intuitive way that fighting strategies in nature vary depending on the value of the resource.

Conventional display does not always occur; sometimes fights are fierce and there is injury or death. Some weapons, such as horns and antlers, have evolved in relation to their efficiency in attack and defence (Geist 1966). In the musk ox, from 5 to 10 per cent of the adult bulls may die each year from fights over females (Wilkinson & Shank

1977) and figures for mule deer indicate that up to 10 per cent of males more than 1.5 years of age show signs of injury each year (Geist 1974). Narwhals (*Monodon monoceros*) use their tusks for fighting and in one study over 60 per cent of the adult males had broken tusks; some had tusk tips embedded in their jaws and most adult males were covered in head scars (Silverman & Dunbar 1980). Smaller animals may also fight viciously. Some male fig wasps have large mandibles which are able to chop another male in half. When several males occur inside the same fig fruit they may engage in lethal combat for the opportunity to mate with females in the fig (Hamilton 1979). One such fruit contained 15 females, 12 uninjured males and 42 damaged males who were dead or dying from fighting injuries. Damage included legs, antennae and heads completely bitten off, holes in the thorax, and eviscerated abdomens.

These are all examples of fights for a valuable resource, namely the opportunity to mate with a female. As we shall see in the next chapter, male–male competition is often intense and many males fail to reproduce altogether. We would expect Hawk-like strategies because failure in a contest could mean failure to pass on genes to future generations; the individuals are really fighting for genetic life or death.

In many contests, however, it is not worth risking serious injury in a fight because the resource is not so valuable. Often failure will not be too disastrous because there will be other opportunities to compete for food or mating. The assumption in the Hawk–Dove game that the costs of serious injury outweigh the value of the resource will probably often hold in nature and so we would expect either a mixture of display and fighting or some conventional settlement. If, as in the Hawk–Dove–Bourgeois game, the resource was sufficiently abundant that individuals found themselves in the role of owner and intruder with about equal probability, then we might predict the Bourgeois strategy would have evolved as the stable way to settle the contest. The rule would be play Hawk if owner and Dove if intruder, so we should observe 'owner wins; intruder retreats'.

Two different examples where the assumptions of the Bourgeois game seem to hold and where the animals do settle their contests in this way are male baboons competing for females in a study in captivity (Kummer, Gotz & Angst 1974) and male speckled wood butterflies (*Pararge aegeria*) competing for mating territories in woodland (Davies 1978a). In both these examples experiments showed that: contests were brief and were always won by the owner; in a contest between two individuals the outcome could be reversed simply depending on who was the owner at the time (Fig. 5.7); when two contestants were tricked experimentally into thinking that they were both the rightful owners, an escalated contest ensued that was damaging to both individuals. In this last experiment, the asymmetry of owner versus intruder was absent and both contestants behaved like Hawks.

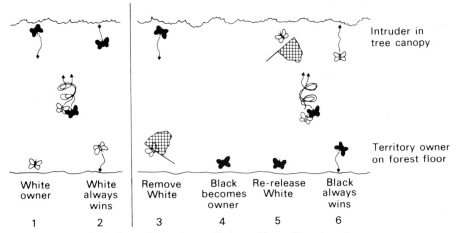

						Intruder in tree canopy
						Territory owner on forest floor
White owner	White always wins	Remove White	Black becomes owner	Re-release White	Black always wins	
1	2	3	4	5	6	

Fig. 5.7 An experiment which shows that male speckled wood butterflies adopt the Bourgeois strategy to settle their contests for territories, i.e. 'owner wins, intruder retreats'. Which of two males wins the contest simply depends on who is the resident (Davies 1978a).

CONTESTS OF STRENGTH

So far we have considered, in a general way, how resource value and the behaviour of other individuals will influence fighting behaviour. However another important factor is the fighting ability or strength of the various contestants. When an individual fights for a resource its decision on how to fight must depend not only on the value of the prize but also on its own strength compared with that of its opponent. In the simple game theory models discussed above we assumed that the costs of injury and display were the same for all fights. This is obviously too simple; animals in a population will differ in their strength and the costs of fighting a large individual will certainly be greater than those of fighting a small one. Therefore we might expect displays to signal fighting ability and so allow the contestants to settle their dispute quickly without recourse to a costly fight (Parker 1974). Furthermore, displays used in assessment should be reliable signals of strength, otherwise weak individuals would be able to mimic the signal and gain an advantage.

In the autumn, red deer stags (*Cervus elaphus*) compete with each other for females. Reproductive success depends on fighting ability; the strongest stags are able to command the largest harems and enjoy the most copulations. Although fighting brings great potential benefits, it also entails serious costs. Almost all males suffer some slight injuries and between 20 and 30 per cent of stags will become permanently injured sometime during their lives, through broken legs or being blinded by an antler point for example. Competing stags minimise fighting costs by assessment of each other's fighting potential and so avoid contests with individuals they are unlikely to beat (Clutton-Brock *et al.* 1979; Clutton-Brock & Albon 1979).

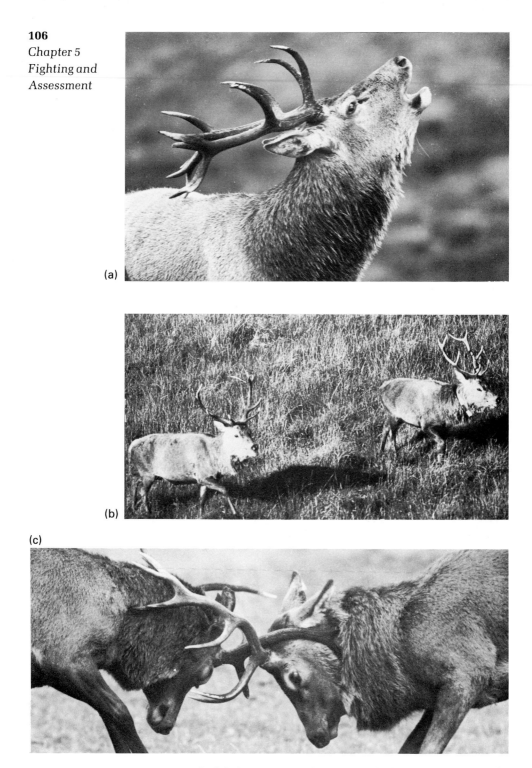

Fig. 5.8 Stages of a fight between two red deer stags. The harem holder roars at the challenger (a). Then the pair engage in a parallel walk (b). Finally they interlock antlers and push against each other (c). Photographs by Tim Clutton-Brock.

In the first stage of the display, harem holder and challenger roar at each other (Fig. 5.8). They start slowly at first and then escalate in rate. If the defender can roar at a faster rate then the intruder usually retreats. Roaring is a good signal of fighting ability because to roar well a stag has to be in good physical condition. At the peak of the rut, harem holders may be roaring at intruders throughout the day and night and, because they do not feed much during this time, they show a steady decline in body weight. Some stags literally become 'rutted out' and are unable to roar strongly. Their females may then be taken over by other stags. In the second stage of a contest if a challenger outroars the defender, or matches him, then he approaches and both stags engage in a parallel walk (Fig. 5.8). This presumably enables them to assess each other more closely. Many fights end at this stage, but if the contestants are still equally matched then a serious fight ensues where they interlock antlers and push against each other. Body weight and skillful footwork are important determinants of victory, but there is a chance that even the winner may get injured. The important point is that these escalated fights are rare and most contests are settled at an earlier stage by displays.

Many animal contests proceed like this and are direct or indirect trials of strength. Buffalos charge at each other and assess their fighting potential in head-on clashes (Sinclair 1977). Beetles engage in pushing contests in which the larger one emerges as victor (Eberhard 1979). Male frogs and toads have wrestling matches where the large ones win the best territories or the most females (chapter 6).

In many species of frogs and toads the pitch of a male's croak is closely related to his body size; the larger the male, the larger the vocal cords and so the deeper the croak. Common toads (*Bufo bufo*) assess the body size, and hence fighting potential, of their rivals by the pitch of their croaks. An attacker is much less likely to attempt a takeover of a female when deep croaks are broadcast from a loudspeaker next to the pair than when high pitched croaks are played (Fig. 5.9).

Often both strength and ownership are cues used to settle contests. When crabs compete for burrows (Hyatt & Salmon 1978) or when spiders compete for webs (Riechert 1978, 1979) the larger individuals usually win the fights, but if the contestants are equal in size then it is often the owner who wins. The rule for deciding the outcome is, 'if larger behave like a Hawk, if smaller like a Dove, if equally matched in strength, adopt the Bourgeois strategy'. In other words our earlier game theory models are too simple because in the real world animals will not play fixed strategies but, rather, will adopt strategies conditional on the fighting potential of their opponents.

FIGHTING FOR DOMINANCE IN FLOCKS

Some animals live in groups and individual differences in fighting ability determine who will have priority of access to food or mates.

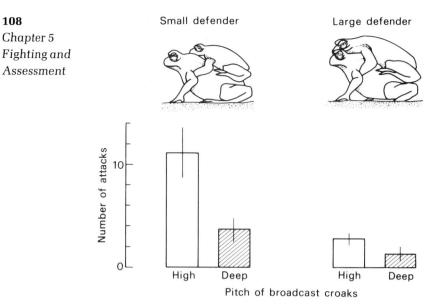

Fig. 5.9 An experiment on fighting assessment in toads, *Bufo bufo*. Medium-sized males attack either small or large paired males, which are silenced by means of a rubber band passing through their mouths. During an attack tape-recorded croaks are broadcast from a loudspeaker next to the pair. For both sizes of defender there are fewer attacks when the deep croaks of a large male are played than when the high-pitched croaks of a small male are broadcast. Therefore croak pitch is used to assess a rival's body size, which is a good predictor of fighting ability because large males are more difficult to displace. Croaks cannot be the only assessment cue, however, because for either croak-pitch there are fewer attacks at large defenders. The strength of a defender's kick may also be important (Davies & Halliday 1978).

Many displays within groups probably involve reliable assessment of strength, however the exact link between the display and fighting ability is not always obvious.

In the Harris sparrow (*Zonotrichia querula*) there is enormous plumage variability in winter when the birds go around in flocks searching for food (Fig. 5.10). Individuals with the blackest plumage are the dominants and they always displace pale birds from food supplies (Rohwer & Rohwer 1978). If blackness signals dominance then why don't the subordinates grow blacker feathers and enjoy a rise in status? In the spring all the birds come into breeding plumage and

Fig. 5.10 Plumage variability in the Harris sparrow. The darker males are dominant in the flocks and win most of the fights.

even the pale males develop black colouration, so the dark feathering is not an obviously uncheatable signal of strength like the roar of a rutting stag or the deep croak of a large male toad.

The Rohwers attempted to create 'cheats' in the flock by experimental treatment of subordinate birds (Table 5.3). In the first experiment they simply painted subordinates black. These treated birds were attacked by others and failed to rise in status. Secondly, some subordinates were injected with testosterone but their pale plumage was left unaltered. These birds behaved more aggressively and attempted to assert their dominance but they did not rise in status because their opponents did not retreat during disputes. Finally, subordinates were painted and injected so that they both looked and behaved like dominants. This time the cheating worked; the birds won more fights and were respected by others in the flock.

Table 5.3 Summary of the experiments on signals of dominance in flocks of the Harris sparrow (after Rohwer & Rohwer 1978).

Experimental treatment of subordinates	Look dominant	Behave as dominant	Rise in status?
1 Paint black	Yes	No	No
2 Inject testosterone	No	Yes	No
3 Paint black and inject with testosterone	Yes	Yes	Yes

Therefore the earlier attempt to create a cheat by painting alone did not fail simply because the other birds didn't like the paint! It failed because although the darkened birds looked dominant, they did not behave so. The conclusion is that plumage alone is not a passport to high status in the flock; the signal must be backed up by dominant behaviour as well. Nevertheless the dark plumage still acts as a badge to reduce the amount of fighting in the flock. In another experiment dominant birds were bleached so that they signalled low rank. They were attacked a lot by others who tried to displace them: they didn't yield however and eventually exerted their superiority but only after a lot of squabbling.

It is clear from the experiments above that cheating is prevented because a subordinate cannot become dominant through darker plumage alone. But why can't a low ranking bird increase its testosterone level as well? This same question arises from work on red grouse (Watson 1970) where males were injected with testosterone and immediately doubled their territory size and increased the number of females they attracted. Presumably the answer to this problem is that an increase in androgen levels would push the behaviour of a subordinate bird beyond its real capabilities. Although it may gain a short term increase in success, perhaps the long term effect would be that all its strength is sapped and it suffers a decrease in fitness (Silverin 1980).

Information transmission and animal contests

During a contest there are two sorts of information that a display could signal to an opponent.

(a) Information about strength

For example 'I am 2 m tall'. This kind of information is often present in animal displays. It is difficult to fake and we have already encountered examples, such as roaring and deep croaks, which are reliable signals of fighting potential.

(b) Information about intentions

For example 'I am going to attack you' or 'I am going to display for exactly 3 min'. This kind of information will be easier to fake and game theory predicts that information about intentions will not be transmitted in displays.

To see this, imagine a population of individuals which convey accurate information about how long they are prepared to display over a resource before retreating. In this population, any individual that sees its opponent announcing a longer display time than its own would benefit by retreating at once. Such a population would not be stable because a lying mutant, which announced a very long display time, would be favoured. Even weak individuals could gain by signalling long display times. Eventually everyone would be lying and it would no longer pay to believe the message. As Maynard Smith (1979) points out, all this really means is that you should not believe what an opponent at poker tells you! (Box 5.1).

Box 5.1 The War of Attrition (from Maynard Smith 1974).
Imagine two animals contesting over a resource (e.g. food or a mate) of value, V. They are equally matched in fighting ability and the contest is won simply by the animal who is prepared to display for the longest time. The longer the animal displays the greater the cost, m, which is assumed to be a linear function of the duration of the display.

Let animal A select a time T_A and animal B select a time T_B. The costs associated with these times are m_A and m_B. If $T_A > T_B$, then we have

Pay-off to A $= V - m_B$
Pay-off to B $= \quad - m_B$.

Note that A only pays the cost of displaying for time T_B because B gives up first.

The problem is how should an animal choose its display time? It is clear that no pure strategy can be an ESS. Any population playing a fixed value, say 'display for 1 min' would

be invaded by a mutant playing a slightly higher value such as
1.1 min. This reasoning might suggest that a contestant should
select the highest possible display time so as to make it difficult
for its opponent to select a higher one. This is not so because if

$$m > \frac{V}{2},$$

it could be invaded by a mutant playing O. (This conclusion
assumes that each contestant pays on average m and wins half
the contests, so that its average gain is

$\frac{V}{2}$. If $m > \frac{V}{2}$,

the net pay-off is less than zero).

It can be shown that the ESS for this game is a mixed one,
namely select the duration of the display according to the
negative exponential distribution.

$$p(x) = \frac{1}{V} \exp. -x/V$$

Where x is the duration of the display and V the gain from
victory. In effect this means that an individual should display
for a random length of time, giving no indication to its
opponent as to the length of time it will display.

Notes
1 See Caryl (1979) for further discussion and applications of
this game to real animal contests. He mentions that the ESS
distribution of display times depends on the relationship
between m and x.
2 In nature we cannot observe directly the distribution of
selected display times that an animal 'bids' in this game (given
by the equation above) because contests are always terminated
by the opponent who plays the lower 'bid'. Parker & Thompson
(1980) derive the expected distribution of contest times we
would observe and test the model on the duration of fights
between male dungflies, *Scatophaga stercoraria*.
3 Most real fights in nature will be settled by asymmetries;
one contestant will be stronger, or the owner, and the other will
quickly retreat. Therefore the assumption of the war of attrition
that the opponents are 'equally matched' and the contest is
symmetrical is unlikely to hold in the real world. Nevertheless
this model is a constructive theoretical exercise because it
shows so clearly that fighting strategies, such as display times,
must be analysed by an ESS approach.

The prediction from this line of argument, that displays should not
evolve to transmit information about intentions, is to a large extent
borne out by the ethological literature. In blue tits (*Parus caeruleus*) a

particular display can be followed by several different types of behaviour and there is no one posture that is followed with a high probability by attack (Stokes 1962; Caryl 1979). In the Siamese fighting fish (*Betta splendens*) there is no consistent difference between the behaviour of the eventual winner and eventual loser until near the end of the contest, so it is impossible to predict which individual will win from its displays (Simpson 1968). The evidence suggests, therefore, that information about intentions is not transmitted (chapter 11).

However there are two lines of research that need further investigation. Firstly, it is clear that animals use a great variety of displays in agonistic encounters and at present game theory has little to say about how the variety might have evolved. Secondly, although it is clear that no one display is a good predictor of future behaviour, sequences of displays may be important. Film analysis of Muhammed Ali in the boxing ring shows that about half of his left jabs are of shorter duration than the visual reaction time. Nevertheless, his opponents still managed to evade most of the punches. They could not have done this by responding to the preliminary stages of the jab. Somehow, therefore, they must have decoded Ali's behavioural sequences so as to predict the punch before it actually starts its movement (Stern 1977).

Summary

The strongest individuals often command the best resources (territories, food, mates) and force others into poor quality habitats. Individuals are expected to assess competitor density and the strength of their rivals when deciding where to settle. Contests are often settled by conventional displays and some simple models show that this will be an ESS when the costs of serious injury in a fight outweigh the benefits arising from victory (speckled wood butterflies). When the pay-off for victory is high there may be fierce fights (fig wasps, narwhals) but contestants often assess each others fighting potential by displays so that the stronger one wins before the fight becomes serious (roaring in deer). Although animals are expected to signal their fighting ability by displays, they are not expected to signal their intentions.

Further reading

Maynard Smith (1979) and Caryl (1980) review the application of game theory to animal contests. Morton (1977) discusses why threat calls tend to be low-pitched. Kroodsma (1979) shows that in wrens the volume of song output seems to be a cue that signals dominance. Yasukawa (1979) describes some neat experiments with junco flocks showing that birds of equal dominance rank may settle contests by the Bourgeois strategy.

Chapter 6. Sexual conflict and sexual selection

Ethologists used to view courtship rituals and mating as harmonious ventures in which male and female co-operated to propagate their respective genes. Admittedly some animals were obviously not co-operative, for example the praying mantis in which the female eats the male during mating, but on the whole courtship was seen as serving functions of common interest to male and female: to 'synchronise the sexual arousal of the sexes', to 'establish the pair bond' to 'allow species identification' and so on. However this view is no longer so widely held and more emphasis is placed on the idea that there are conflicts of interest between male and female in courtship and mating. The sexes are seen as forming an uneasy alliance in which each attempts to maximise its own success at propagating genes. They co-operate because both pass on their genes via the same progeny and therefore each has a 50 per cent stake in the survival of the offspring. But choice of mating partner, provisioning of the zygote with food, and caring for the eggs and young are all issues over which the sexes may conflict. The outcome of this sexual conflict is often more akin to exploitation by one sex of the other than to mutual co-operation.

In order to understand why sexual reproduction should be viewed in this way, we have to go right back to the beginning, to the fundamental difference between male and female.

Males and females

Sexual reproduction entails gamete formation by meiosis and the fusion of genetic material from two individuals. It almost always, but not invariably, involves two sexes called male and female. In higher animals the sexes are often most readily distinguished by external features such as genitalia, plumage, size, or colour, but these are not fundamental differences. In all plants and animals the basic difference between the sexes is the size of their gametes; females produce large, immobile, food-rich gametes called eggs, while male gametes or sperm are tiny, mobile, and consist of little more than a piece of self-propelled DNA. Sexual reproduction without males and females occurs in many protists such as *Paramecium* where the 'gametes' which fuse during sex are of the same size. This is referred to as *isogamous* sexual reproduction. The fusion of two gametes of unequal size, one large and one small is, however, much commoner and occurs in virtually all sexually reproducing multicellular plants and animals. It is called *anisogamous* sex.

As we shall show in the rest of this chapter, the fundamental asymmetry in gamete size has far-reaching consequences for sexual behaviour. Because females put more resources than do males in each zygote male courtship and mating behaviour is to a large extent directed towards competing for and exploiting female investment. However, before we go on to consider some of the consequences of anisogamy, we will look briefly at how it might have evolved in the first place.

THE ORIGIN OF ANISOGAMY

Isogamous sex occurs today in simple unicellular organisms and is generally regarded as the ancestral form of sex. Parker, Baker & Smith (1972) have developed an ingenious and persuasive hypothesis to account for the way in which anisogamy might have evolved from isogamy. They start with the assumption that the survival of a zygote depends on its size; the larger the zygote the more food it has to keep it going during its development and the better its chances of surviving. It is very obvious that the early development and survival of a chicken for example, depends on food reserves inside the hen's egg, and the same is true to a greater or lesser extent in all multicellular organisms. Now suppose that in the ancestral isogamous population there was some genetic variation in gamete size. Larger than average gametes would give rise to larger than average zygotes which would be better at surviving. Large gametes would have been favoured by selection if this increase in survival more than compensated for the fact that large gametes could be made in smaller numbers, given a fixed supply of starting material. For example if the store of reserves could be divided into two large or four small gametes, the zygote produced by large gametes would have to be more than twice as good at surviving in order for large gametes to be favoured by selection.

If large gametes are favoured by selection there is immediately selection on small gametes to find a large partner, fuse with it, and in effect parasitise its food reserves. Even very tiny gametes which would have no chance of producing a viable zygote if they fused among themselves could be successful if they parasitised large gametes. These tiny gametes will be under very strong selection indeed to seek out and mate with a large partner. At the same time large gametes will be selected to resist the small ones, since the most viable zygote of all is one resulting from the fusion of two large cells. But the penalty for failing to resist a small gamete is not as great as the penalty paid by the small one if it fails to find a large partner; a medium sized zygote has some chance of surviving, a small one has none. This means that selection acts more strongly on the small than on the large gametes. The fact that small gametes are, as explained earlier, made in larger numbers also favours their success in exploiting large gametes. Because small ones are likely to occur in a greater variety of genotypes and suffer a higher rate of mortality (both these are consequences of numerical superiority) they will evolve faster and during evolutionary time 'outwit' the defences of large gametes.

Parker, Baker and Smith suggest, therefore, that there was an ancient evolutionary race in which small gametes (sperms) succeeded in parasitising large ones (eggs). During the race anisogamous sex gradually replaced isogamous sex. Sperm have evolved towards greater mobility for seeking out eggs and contain virtually no food reserves, while eggs have given up mobility (and therefore the chance of choosing a partner) and concentrate exclusively on storing food. So extreme is the development of this trend in some species that it has produced the largest single cell in the animal kingdom, the egg of an ostrich.

What, it might be asked, happened to the intermediate sized gametes in the ancestral population? They lost out in the evolutionary race because they neither enjoyed the advantage of great numbers nor the security of large food reserves.

Although Parker *et al.*'s hypothesis cannot be tested directly since it is about evolutionary history, one of its basic assumptions, namely that anisogamy evolved because zygote survival was related to size, is supported by a comparative survey of the algae. Unicellular algae, in which food reserves for growth in the zygote are relatively unimportant, tend to be isogamous, while multicellular genera, where food reserves in the zygote play a role in survival are mostly anisogamous (Table 6.1).

Table 6.1 Within the family of algae called Volvocales, unicellular genera tend to be isogamous while multicellular genera are usually anisogamous. Intermediate sized types are slightly anisogamous. These results support the assumption of Parker *et al.*'s (1972) hypothesis of the origin of anisogamy (Knowlton 1974). Other families of algae show similar, but less clear cut trends (Bell 1978).

Genus	Size of adult (no of cells)	Gametes
Lobomas	Unicellular	Isogamous
Sphaerella	Unicellular	Isogamous
Polytomella	Unicellular	Isogamous
Pandorina	16	Slight anisogamy
Volvox	20000	Anisogamy
Peleodorina	128	Anisogamy

Note
The table gives only a few representative genera.

The assumption that anisogamy has evolved from isogamy has been questioned by Alexander and Borgia (1979). They point out that sexual reproduction in some present day bacteria and Protista (e.g. ciliates, diatoms) involves exchange of micronuclei consisting of genetic material with little or no cytoplasm. The micronucleus, they suggest, is analogous to sperm, while the cell of the organism itself is equivalent to an egg. If, as Alexander and Borgia hypothesise, exchange of micronuclei is an ancient way of reproducing sexually, anisogamy could have preceded isogamy in evolution. But why should the exchange of micronuclei involve donation of genes without cytoplasm? One suggestion is that the donor has no control over the use of resources after transferring them to the receiver. If the donor transferred genes packed in a rich food supply, the way would be open for the recipient to accept the resources and reject the genes. Instead the donor transfers genes without food, and the recipient accepts genes only if it is allowed to donate some in return. It is not known whether the exchange of genes in unicellular organisms is regulated to prevent 'cheating' (donating genes and refusing to accept them) but in a parallel situation, transfer of sperm between simultaneously hermaphroditic fish, each partner accepts only a small quantity of sperm at a time, giving some of its own sperm to the other fish in between bouts of receiving (Fischer 1980). This extraordinary mating ritual has the appearance of a system which has evolved to prevent cheating.

FEMALES AS A SCARCE RESOURCE

Anisogamous sexual reproduction, then, involves parasitism of a large egg by a small sperm. Females produce relatively few large gametes and males produce many small ones. Because of this males can potentially fertilise eggs at a faster rate than they are produced (illustrated by the fact that 5 ml of human semen contain enough sperm to fertilise in theory eggs amounting to twice the population of the U.S.A.), and females are therefore a *scarce resource* for which males compete. A male can increase its reproductive success by finding and fertilising many different females, but a female can only increase her success by turning food into eggs or offspring at a faster rate. This point is graphically demonstrated by mammals such as man, in which a female spends many months producing a single child, during which time a male could potentially fertilise hundreds of other mates. Only by speeding up her production of young can the female have more children in a lifetime. The same argument holds whenever females invest more than males in each offspring, whether the investment is in the form of food in the egg or care of the eggs and young later on.

This point was neatly summarised by Robert Trivers (1972), who was the first person to emphasise the relationship between the investment of resources in gametes and other forms of care, and sexual competition. He wrote 'Where one sex invests considerably more than the other, members of the latter will compete among themselves to mate with members of the former'. The term 'investment' was used by Trivers to refer to the effort put into rearing an individual offspring from the parent's limited pool of resources. The sum of parental investment in all offspring during a parent's lifetime is referred to as

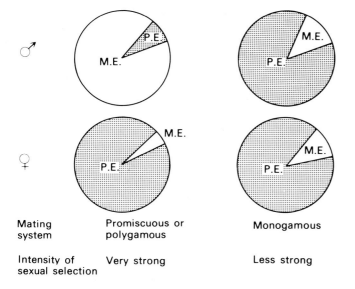

	Promiscuous or polygamous	Monogamous
Mating system		
Intensity of sexual selection	Very strong	Less strong

Fig. 6.1 The total resources of time and energy used by an animal in reproduction is referred to as reproductive effort. This is represented by a circle. Reproductive effort can be partitioned into parental effort (provisioning and rearing offspring) and mating effort (acquiring mates). These are represented by the shaded and open areas of the circles respectively. In general males put relatively more into mating effort than do females, but this varies between species. The intensity of sexual selection therefore also varies. The differences in relative parental effort of the sexes is often related to the mating system. In monogamous species male and female effort is more similar than in polygamous and promiscuous species (see chapter 7) (after Alexander & Borgia 1979).

'parental effort'. Females generally put most of their reproductive effort into 'parental effort' while males put most of theirs into 'mating effort' (Fig. 6.1).

The consequence is that males have a much greater reproductive potential than females (Table 6.2) and are therefore under strong selection to be good at seeking out and competing for females: the payoff for a successful male in terms of offspring fathered is enormous. Much of male reproductive behaviour can be understood with these ideas in mind.

Table 6.2 Two examples to illustrate the point that males in most species have a much higher potential reproductive rate than females. The data for man come from the Guinness Book of Records. The man was Moulay Ismail the Bloodthirsty, Emperor of Morocco, and the woman produced her children in 27 pregnancies. The data for elephant seals are from the work of Le Boeuf (1974).

Species	Maximum no. of offspring per lifetime	
	Male	Female
Elephant seal	200	c.15
Man	888	69

The different factors limiting male and female reproductive success were first demonstrated experimentally by A. J. Bateman (1948). He put equal numbers of male and female fruit flies (*Drosophila*) in a bottle and scored the number of matings and offspring produced by each individual using genetic markers to assign parentage. Bateman demonstrated the two main points we discussed in the preceding paragraphs. First, the reproductive success of a male depends on the number of females he fertilises while for females a single copulation is enough to achieve more or less maximum success. In other words it pays males, but not females to seek out more matings. The second point is that some males do extremely well and others do very badly in terms of fathering offspring, while females all do about equally well. Male reproductive success has a higher variance because of competition between males for mates; some individuals are successful and others are not.

THE SEX RATIO

If one male can fertilise the eggs of dozens of females why not produce a sex ratio of, say, one male for every 20 females? With this ratio the reproductive success of the population would be higher than with a 1:1 ratio since there would be more eggs around to fertilise. Yet in nature the ratio is usually very close to 1:1 even when males do nothing but fertilise the female. As we saw in chapter 1, the adaptive value of traits should not be viewed as being 'for the good of the population', but 'for the good of the individual', or more precisely 'for the good of the gene'. As R. A. Fisher (1930) first realised the 1:1 sex ratio can readily be explained in terms of selection acting on the individual; his argument is simple but subtle.

Suppose a population contained 20 females for every male. Every male has 20 times the expected reproductive success of a female (because there are on average 20 mates per male) and therefore a parent whose children are exclusively sons can expect to have almost 20 times the number of grandchildren produced by a parent with mainly female offspring. A female biassed sex ratio is therefore not evolutionarily stable because a gene which causes parents to bias the sex ratio of their offspring towards males would rapidly spread, and the sex ratio will gradually shift towards a greater proportion of males than the initial 1 in 20. But now imagine the converse. If males are 20 times as common as females a parent producing only daughters will be at an advantage. Since one sperm fertilises each egg, only one in every 20 males can contribute genes to any individual offspring and females therefore have twenty times the average reproductive success of a male. So a male biassed sex ratio is not stable either. The conclusion is that the rarer sex always has an advantage, and parents which concentrate on producing offspring of the rare sex will therefore be favoured by selection. Only when the sex ratio is exactly 1:1 will the expected success of a male and a female be equal and the population stable. Even a tiny bias favours the rarer sex: in a population of 51 females and 49 males where each female has one child, an average male has 51/49 children. This *average* value is the same whether one male does most of the fathering or whether fatherhood is spread equally among the males.

The argument that the sex ratio should be 1:1 can be refined by re-phrasing it in terms of investment. Suppose sons are twice as costly as daughters to produce because for example, they are twice as big and need twice as much food during development. When the sex ratio is 1:1 a son has the same average number of children as a daughter. But since sons are twice as costly to make they are a bad investment for a parent: each of its grandchildren produced by a son is twice as costly as one produced by a daughter. It would therefore pay the parents to concentrate on making daughters. As the sex ratio swings towards a female bias, the expected reproductive success of a son goes up until at a ratio of two females to every male each son produces twice the number of children produced by a daughter. At this point sons and daughters give exactly the same return per unit investment; a son costs twice as much to make but yields twice the return. This means that when sons and daughters cost different amounts to make, the stable strategy in evolution is for the parent to *invest* equally in the two sexes and not to produce equal numbers (Box 6.1).

Box 6.1 Tests of Fisher's theory of the sex ratio

It is commonly found that the sex ratio in species with equal sized males and females is 1:1. However this is not a very convincing test of the theory since it could be argued that a 1:1 ratio is a by-product of the mechanism of meiosis, in which there is an equal chance that a zygote will contain male or female sex chromosomes. More convincing tests of the theory are observations of deviations from the 1:1 ratio which accord with predictions:

1 The sex ratio is biassed towards females when there is local mate competition

Jack Werren (1980) has tested the prediction that the degree of bias depends on the extent of local mate competition. He studied the parasitoid wasp *Nasonia vitripennis* which lays its eggs inside the pupae of flies such as *Sarcophaga bullata*. If one female parasitises a pupa, her daughters are all fertilised by her sons and as predicted the sex ratio of her clutch of eggs is biassed towards females. Only 8.7 per cent of the brood is male. If a second female lays her eggs in the same pupa, what should her sex ratio be? If she lays few eggs she should produce mainly sons, since the first female has laid predominantly female eggs. But as the proportion of the total number of eggs in the pupa that come from the second female increases, the chance that sons of the second female will compete for mates also increases. Therefore her brood should have a female biassed sex ratio. Werren found exactly this pattern: when the second female's clutch was 1/10 the size of the first female's it contained only males, but when it was twice as large as the first female's it contained only 10 per cent males.

2 When the costs of making males and females differ, investment in the sexes is equalised by biassing the sex ratio. Bob Metcalf (1980) studied the sex ratio of two species of paper wasp *Polistes metricus* and *P. variatus*. In the former females are smaller than males, while in the latter the two sexes are similar in size. As predicted, the population sex ratio is female biassed in *metricus* and not in *variatus*. The result is that in both species the investment ratio is 1:1.

3 When the population ratio of investment deviates from 1:1, a compensatory bias in favour of the rarer sex should occur. In Metcalf's study of *P. metricus* he found that some nests produced only male offspring. As explained in chapter 10 these offspring are the product of unfertilised eggs and are produced by workers in nest where the queen has died. In the remainder of the nests in the population Metcalf found a female biassed sex ration, so that the ratio of investment in the population as a whole is 1:1.

It is important to note that these tests of Fisher's theory all refer to Hymenoptera, in which the female can control the sex of her offspring by whether or not she fertilises her eggs. (Unfertilised eggs become males, fertilised eggs develop into females see chapter 10.) In species in which sex determination depends on sex chromosomes there is little evidence that the ratio can vary from 1:1. Artificial selection with domestic animals has failed to bias the sex ratio towards females and this suggests that there may be little or no genetic variability in the sex ratio in many animals. This means that the adaptive modifications of sex ratio described in this box may not occur outside the Hymenoptera and other groups with similar mechanisms of sex determination (Williams 1979). The implications of this are discussed by Maynard Smith (1980).

Fisher's theory predicts a different outcome when brothers compete with each other for mates (so-called 'local mate competition'). Suppose, for example, that two sons have only once chance to mate and that they compete for the same female. Only one of them can be successful in mating, so from their mother's point of view one of them is 'wasted'. This is an extreme example, but it illustrates the general point that when sons compete for mates their value to their mother is reduced. The mother should therefore bias her ratio of investment towards daughters. The exact degree of bias predicted by Fisher's theory depends on the degree of local mate competition. Extreme competition is to be expected in species with limited powers of dispersal (because brothers will stay together in the same place) and therefore it tends to be associated with inbreeding. In the extreme case of inbreeding, a mother 'knows' that all her daughters will be fertilised by her sons. The best sex ratio in this instance is to produce just enough sons to fertilise the daughters, since any other males will be wasted. The crucial difference between this and the earlier argument for a 1:1 sex ratio is that here the ratio of males to females in the rest of the population does not matter. A female biased ratio within a brood will not give other parents a chance to benefit by concentrating on sons. An example which supports this prediction is the viviparous mite *Acarophenox* which has a brood of one son and up to 20 daughters. The male mates with his sisters inside the mother and dies before he is born (Hamilton 1967).

Finally it is worth pointing out that the theory of sex ratios discussed here is one example of a more general theory of *sex allocation*. Other examples of the problem of allocation of resources to male and female reproduction include the division of resources into eggs and sperms by simultaneous hermaphrodites and the timing of sex change in sequential hermaphrodites (see chapter 8).

Sexual selection

The combination of gamete dimorphism and 1:1 population sex ratio means that males usually compete for females. The potential pay-off for male success is high, so selection for male ability to acquire matings is very strong. This kind of selection is usually referred to as sexual selection. It can work in two ways: by favouring the ability of males to compete directly with one another for fertilisations, for example by fighting (*intra-sexual selection*), or by favouring traits in males which attract females (*inter-sexual selection*). Often the two kinds of selection act at the same time.

The intensity of sexual selection depends on the degree of competition for mates. This in turn depends on two factors: the difference in parental effort between the sexes and the ratio of males to females available for mating at any one time (referred to as the operational sex ratio). When parental effort is more or less equal, as for example in monogamous birds where both male and female feed the young, sexual selection is less intense than in species with very different levels of parental effort (Fig. 6.1). This follows from the point made earlier (p. 116) that the sex making little investment competes for the sex which makes a large investment. If equal numbers of the two sexes come into breeding condition at the same time, the degree of sexual selection is reduced because there is less chance for a few males to control access to very large numbers of females. In contrast, when females come into breeding condition asynchronously there is a chance for a small number of males to control many females one after the other. With such high potential pay-offs, sexual competition is very intense (see chapter 7).

ARDENT MALES

The most dramatic and obvious way in which males compete for mates is by fighting and ritualised contests. Males may dispute over direct access to females or over places where females are likely to go, as for example when male speckled wood butterflies defend patches of sunlight (chapter 5). Fighting is often a risky business, as illustrated by the injuries sustained by red deer stags referred to in chapter 5. The most intense fights in many species occur when females are ready to be fertilised and once a male finds a female he often guards her (Fig. 6.2).

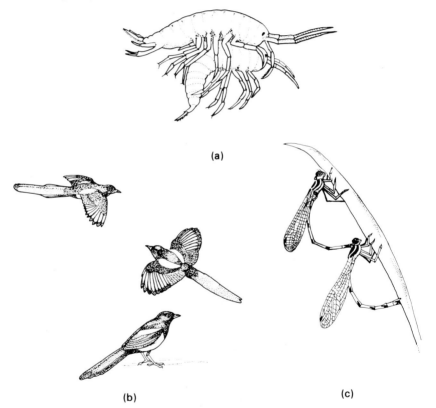

(a)

(b)   (c)

Fig. 6.2 Mate guarding as a form of sexual competition. (a) Precopulatory mate guarding in the freshwater amphipod *Gammarus*. The mature female in this species is ready to be fertilised immediately before she moults. Males guard females in the few days preceding this moult (Birkhead & Clarkson 1980). (b) Male European magpies (*Pica pica*) assiduously guard their mates against intruding males just before and during the period of egg laying (Birkhead 1979). (c) After copulation the male damselfly guards the female while she lays her eggs, by clasping her thorax with the tip of his abdomen in the 'tandem' position (Corbet 1962).

Males often compete in ways which are less conspicuous than fights, but are no less effective and often more bizarre. The invertebrates are a particularly rich seam of examples. Females of the dragonfly *Calopteryx maculata*, as with many other insects, mate with a number of males and store the sperm in a special sac in the body for use at a later date. The males compete for fertilisations by trying to ensure that previous sperm is not used by the female. The penis of a male *Calopteryx* is equipped with two special scoops at the end which are used to scrape out of the female any sperm left by previous males before he injects new sperm into the sperm sac (Fig. 6.3; Waage 1979).

In some invertebrates (especially insects) the male cements up the female's genital opening after copulation to prevent other males from fertilising her. The males of *Moniliformes dubius*, a parasitic acanthocephalan worm in the intestine of rats, produces a chastity belt of this kind but in addition to sealing up the female after copulation, the male sometimes 'rapes' rival males and applies cement to their genital region to prevent them from mating again (Abele & Gilchrist 1977). No less remarkable are the habits of the hemipteran insect *Xylocoris maculipennis*. In normal copulation of the species the male simply pierces the body wall of the female and injects sperm, which then swim around inside the female until they encounter and fertilise her eggs. As with the acanthocephalan worms, males sometimes engage in homosexual 'rape'. A male *Xylocoris* may inject his sperm into a rival male. The sperms then swim inside the body to the victim's testes, where they wait to be passed on to a female next time the victim mates (Carayon 1974).

Fig. 6.3 The penis of a male damselfly (*Calopteryx*), showing the sperm mass of a previous male (sm) attached to the projections on the penis (Photo by J. K. Waage).

Competition between males to prevent each other's sperm from fertilising eggs is sometimes referred to as 'sperm competition'. Another insect example was described in chapter 3: the sperm of a second male displaces that of the first male to mate with a female dungfly. Sperm competition also occurs in vertebrates. For example during courtship male salamanders and newts deposit little sperm-capped rods of jelly (spermatophores) on the bottom of the pond and then try to manouever the female onto the spermatophore to achieve fertilisation. In the salamander *Ambystoma maculatum* males compete by depositing their spermatophores on top of those of other males. The top spermatophore is the one that fertilises the female's eggs (Arnold 1976).

A fourth example of the arcane methods of male-male competition found among invertebrates is the use of anti-aphrodisiac smells. Larry Gilbert (1976) noticed that female *Heliconius erato* butterflies always smell peculiar after they have mated. He was able to show experimentally that the scent does not come from the female herself, but is deposited by the male at the end of mating. Gilbert also found that the scent discourages other males from mating with the female, perhaps because it resembles a scent used by males to repel one another in other contexts.

RELUCTANT FEMALES

Since females in the great majority of species are the chief providers of resources for the zygote, they might be expected choose their mates carefully in order to get something in return. To put it another way, each egg represents a relatively large proportion of a female's lifetime production of gametes when compared with a sperm, so the female has more to lose if something goes wrong. Mating with the wrong species could cost a female frog her whole year's supply of eggs, but would cost the male very little apart from lost time—he could still go on to mate successfully with a member of the correct species the next day. Not surprisingly therefore, females are on the whole choosier than males during courtship. Choosiness extends not only to discriminating between species, but also to discriminating between males within a species. Females often select males on the basis of material resources they can offer and perhaps sometimes to obtain good genes for their offspring.

(a) Good resources

In many animal species males defend breeding territories containing resources which play a crucial role in the survival of a female's eggs or young (see also chapter 7). For example male North American bullfrogs (*Rana catesbiana*) defend territories in ponds and small lakes where females come to lay their eggs (Fig. 6.4). Some territories are much better for survival of eggs than others and these are the ones which females prefer. One factor which has an important influence on survival of eggs is predation by leeches (*Macrobdella decora*). Two environmental features of a territory influence leech predation: if the water is warm the eggs develop faster and are therefore exposed to predation for fewer days, and if the vegetation in the water is not too dense the eggs can form into a ball which the leeches find hard to attack. In territories with a dense mat of vegetation the eggs lie in a thin film on top of the plants and are more easily attacked. The bullfrogs

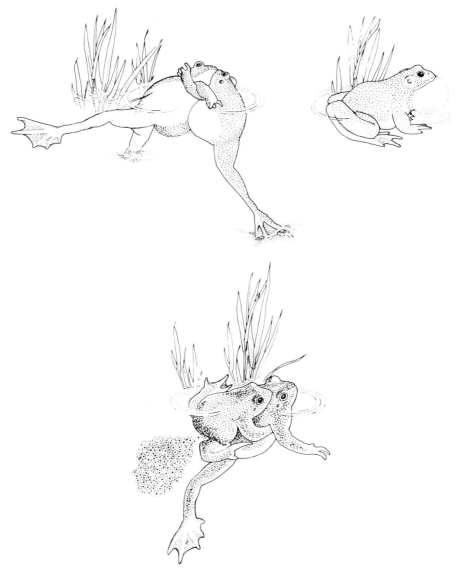

Fig. 6.4 Sexual selection in male bullfrogs. Males compete by wrestling and calling (top) for good territories, in which the females prefer to lay their eggs (bottom). The good territories have high survival of eggs because they are warm and because the vegetation is not too dense (Howard 1978a, b).

also show that female choice and male–male competition may go hand in hand. The preferred territories are hotly contested by males and the largest, strongest frogs end up in the best sites.

Food is a resource which often limits a female's capacity to make eggs and during courtship females may choose whether or not to mate with a male on the basis of his ability to provide food. In some birds and insects for example, males may provide food for the female during courtship ('courtship feeding') which makes a significant contribution to her eggs. Female hanging flies (*Hylobittacus apicalis*) will mate

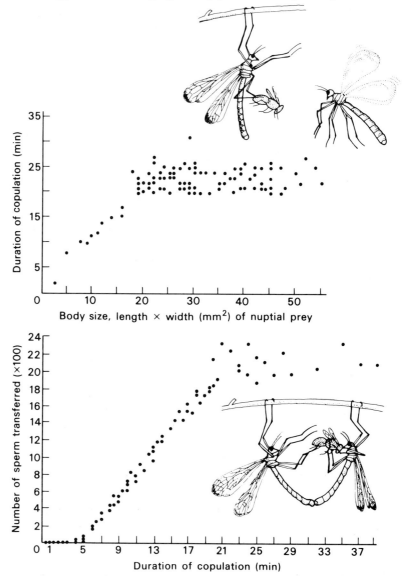

Fig. 6.5 Female choice for good resources. Female hanging flies (*Hylobittacus apicalis*) mate with males for longer if the male brings a larger prey item to eat during copulation. The male benefits from long copulation because he fertilises more eggs (Thornhill 1976).

with a male only if he provides a large insect for her to eat during copulation. The larger the insect, the longer the male is allowed to copulate and the more eggs he fertilises (Fig. 6.5). The female gains from a large insect by having more food to put into her eggs. In birds, the male usually helps to feed the young and courtship feeding may play the additional role of indicating to the female how good the male is at bringing food for the young. In the common tern (*Sterna fuscata*) there is a correlation between the ability of the male to bring food during courtship feeding and his ability to feed the chicks later in the season. Pairs often break up during the courtship feeding period and it is possible that females are assessing their mates and rejecting poor partners (Nisbet 1977).

(b) Good genes

If some males have 'better' genes than others, could a female improve the success of her progeny by choosing males with good genes? Good genes are ones which increase the ability of her offspring to survive, compete and reproduce. One of the very few studies in which this idea has been tested is a clever experiment by Linda Partridge (1980). She took groups of female fruit flies (*Drosophila*) and either allowed them to mate freely with a population of males or forced each female to mate with a randomly chosen partner. The offspring of the 'choice' and 'no choice' females were then tested for their competitive ability by rearing the larvae in bottles with a fixed number of standard competitors (these were distinguishable by a genetic marker). Partridge found that the offspring of the 'choice' group did slightly but consistently better than those of 'no choice' females in the larval competition experiments. Therefore by choosing mates, females can increase the survival of their offspring and since male fruitflies contribute only DNA to their offspring, the females must somehow be able to choose good genes. It is not known why the chosen genes are good, although one possibility is that females can select males with different genotypes from their own and confer heterozygous advantage on their offspring. This advantage would not however be passed on to the next generation since heterozygotes do not breed true.

The theory of sexual selection is most famous as an attempt to explain the evolution of excessively elaborate adornments and displays of male peacocks, pheasants, birds of paradise and so on (Fig. 6.6). Some elaborate displays may have evolved for use in contests between males, but the most widely accepted (and still controversial) account of how such traits evolve depends on selection by females for good genes. In this case, however the genes are not ones which increase the ability of a female's offspring to survive and compete (in fact the bizarre adornments usually seem to be a hindrance to survival) but rather for genes which increase the sexual attractiveness of her sons.

R. A. Fisher (1930) was the first to clearly formulate the idea that elaborate male plumage may be sexually selected simply because it makes males attractive to females. This may sound circular, and indeed it is, but that is the elegance of Fisher's argument. At the beginning, he supposed, females preferred a particular male trait (let us take long tails as an example) because it indicated something about male quality. Perhaps males with longer tails were better at flying and

Fig. 6.6 The exotic plumes of the greater bird of paradise (*Paradisea apoda*) are good examples of male displays which might have evolved as a result of female choice. The male is on the right, the female on the left.

therefore collecting food or avoiding predators. (An alternative starting point is to suppose that longer tails were simply easier for females to detect and that they therefore attracted more mates). If there is some genetic basis for differences between males in tail length the advantage will be passed on to the female's sons. At the same time, a gene which causes females to prefer longer than average tails will also be favoured since these females will have sons better able to fly. Now once the female preference for longer tails starts to spread, longer tailed males will gain a double advantage: they will be better at flying *and* be more likely to get a mate. The female similarly gets a double advantage from choosing: she will have sons that are both good fliers and attractive to females. As the positive feedback between female preference and longer tails develops, gradually the benefit of attractive sons will become the more important reason for female choice, and the favoured trait might eventually decrease the survival ability of males. When the decrease in survival counter-balances sexual attractiveness, selection for increasing tail length will grind to a halt.

This simple but logical account of how extreme traits could be sexually selected has been criticised by Amotz Zahavi (1975, 1977). He points out that the long tail of the peacock is a handicap in day to day survival, a fact which few would dispute. He then goes on to suggest that longer tails are preferred by females precisely *because* they are a handicap and therefore act as an advertisement of male quality. The tail demonstrates a male's ability to survive in spite of the handicap, which means that he must be extra good in other respects and is able to pass on any genetic component of this quality to his sons and daughters. The difficulty with this argument is that the offspring inherit not only the 'good genes' but also the handicap, and under most conditions they will be better off without either than with both.

There are surprisingly few studies to test whether supposed sexually selected traits such as peacock's tails are preferred by present day females. One such study was done on the European sedge warbler (*Acrocephalus scirpaeus*) in which the male has an extraordinarily elaborate song described by Clive Catchpole (1980) as an 'acoustic peacock's tail'. The song consists of a long stream of almost endlessly varying trills, whistles and buzzes and is sung by the male after arriving back on the breeding territory from the winter quarters: as soon as a male pairs, it stops singing. The sedge warbler is a monogamous bird and therefore sexual selection might be expected to be less intense than in many other species (Fig. 6.1). However the sedge warbler is a migrant with a fairly short breeding season, so there may well be intense competition among males to acquire as mates the first females to arrive on the breeding grounds. In a variety of seasonally breeding birds it has been found that the earliest pairs to breed are the most successful, probably because they produce their young while food is still at its seasonal peak of abundance (Perrins 1970). If this generalisation applies to sedge warblers there would be a considerable advantage for males that successfully attracted the first females.

Catchpole's measurements showed that the males with the most elaborate songs are the first to acquire mates (Fig. 6.7), as expected from Fisher's hypothesis of sexual attractiveness.

Perhaps the most controversial aspect of the idea that females choose males for good genes is the problem of genetic variance. This can be illustrated with an hypothetical example. Suppose a farmer wants to select for larger body size in a population of turkeys. He takes the heaviest males and females to start the next brood and repeats this procedure for several generations. What will happen? Assuming that there is some genetic variance for body weight, selection will at first be fairly effective, but soon the stock will become less variable with respect to genes for body weight, because only a few genotypes (the heaviest) have been allowed to breed. When the genetic variance is 'used up' selection will cease to be effective in changing body size. In the same way, females cannot improve the genetic quality of their

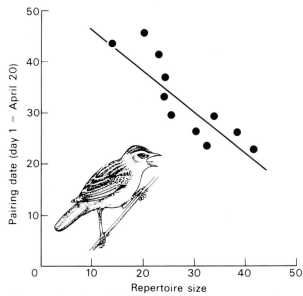

Fig. 6.7 Male sedge warblers with the largest song repertoires are the first to acquire mates in the spring. The size of song repertoire is estimated from sample tape recordings of each male. The results were collected in such a way as to control for the possibilities that older males, or males in better territories, both mate first and have larger repertoires (Catchpole 1980).

offspring indefinitely by choosing males for good genes. So there is something of a paradox. On the one hand female fruit flies seem to be able to select for good genes and the only accepted explanation of how very elaborate plumages and displays evolve depends on selection of good genes, but on the other hand the simple theories of population genetics should make us sceptical. Probably the genetic theory is too simple. Genetic variation for fitness-related traits might well be maintained in natural populations either because of new mutations or, more likely, because the optimal genotype is different in different places or at different times so that selection does not favour the same genes for long enough to exhaust all the genetic variation.

MALE INVESTMENT

We have so far assumed that females are investors and males are competitors. While this picture describes most animal species, there are exceptions. In many birds, some amphibians, fish, and arthropods both male and female invest about equally in the eggs or young by feeding, guarding or brooding. Sometimes the usual sex roles are completely reversed so that males do the investing and females the competing (chapter 7). The ideas about sexual conflict and sexual selection can still be applied in modified form to species with equal or primarily male investment. When both sexes care equally for the offspring, for example, courtship may involve assessment and choice by males as well as by females. Males of species with internal fertilisation can never be absolutely sure that they have fathered the children of their mate, and one role of courtship may be as an insurance against cuckoldry. A prediction of this idea is that courtship allows males to assess whether or not females have previously mated with others. This was tested by Erickson & Zenone (1976). They found that male barbary doves (*Streptopelia risoria*) attack a female instead of courting her if she performs the 'bow posture' (an advanced stage of courtship) too quickly. Since the females which responded in this way had been pre-treated by allowing them to court with another male, the reaction of the test males in rejecting eager females is adaptive if courtship plays a role in assessing certainty of paternity, before investing in offspring. It would not have been predicted by the older view that male courtship serves to sexually arouse the female!

Sexual conflict

Let us now return to the starting point of this chapter, sexual conflict. Recall that Parker *et al.* view the origin of anisogamy as the primeval example of sexual conflict. The conflict was one about mating decisions. Macrogametes might have done better had they been able to discriminate against microgametes, but in the evolutionary race microgametes won. Similar, but more directly observable, conflicts of interest between the sexes are still apparent today, not only with respect to mating decisions but also in the contexts of parental investment, multiple matings and infanticide.

(a) Mating decisions

As we have emphasised earlier in the chapter, females have more to lose and therefore tend to be choosier than males. Thus for a given encounter it will often be the case that males are favoured if they do mate and females if they do not (Parker 1979). An extreme manifestation of this conflict is enforced copulation or 'rape', as exemplified by scorpionflies (*Panorpa* spp.). Male scorpionflies usually acquire a mate by presenting her with a nuptial gift in the form of a special salivary secretion or a dead insect (this is very similar to the *Hylobittacus* described earlier). The female feeds on the gift during copulation and turns the food into eggs. However, sometimes a male copulates by raping the female: he grasps her with a special abdominal organ (the notal organ) without offering a gift (Thornhill 1980). Rape appears to be a case of sexual conflict. The female loses because she obtains no food for her eggs and has to search for food herself, while the male benefits because he avoids the risky business of finding a nuptial gift. Scorpionflies feed on insects in spiders' webs and quite often get caught up in the web themselves, so foraging is certainly risky (65 per cent of adults die this way). Why do not all males become rapists? The exact balance of costs and benefits is not known, but it appears that rapists have a very low success rate in fertilising females, so perhaps males adopt this strategy only when they cannot find prey or make enough saliva to attract a female.

(b) Parental investment

This is a topic to which we will return in the next chapter. Here it is sufficient to note that in species with investment beyond the gamete stage, each sex might be expected to exploit the other by reducing its own share in the investment. The outcome of this sexual conflict may depend on practical considerations such as which sex is the first to be in a position to desert the other. When fertilisation is internal, for example, a male has the possibility of deserting the female immediately after fertilisation and leaving her with the eggs or young to care for.

(c) Infanticide

As we saw in chapter 1, male lions may slaughter the cubs in a pride shortly after they take over as group leaders. This behaviour (which is also seen in some primates) probably increases male reproductive success, as explained in chapter 1, and clearly decreases female success. This seems to be a case of sexual conflict in which the males have won, but it is perhaps surprising that females have not evolved counter-adaptations. They could, for example eat their own young once they have been killed in order to recoup as much as possible of their losses.

(d) Multiple matings

As Bateman's experiments with *Drosophila* showed (p. 118) females often gain little by mating with more than one male. However, because of sperm competition males may gain by mating with already fertilised females. Multiple matings are likely to be costly to the female at the same time as being advantageous to the male. This is dramatically illustrated by the dungflies described in chapter 3. When two males struggle for possession of a female, the female is sometimes drowned in cowdung by the fighting males on top of her!

Conflicts of interest between the sexes will lead to an evolutionary race of the sort envisaged by Parker *et al.* for sperms and eggs. There is no simple answer to the questions 'Which sex is more likely to win the chase?' As we discussed earlier, factors such as the strength of selection and the amount of genetic variation will determine how fast the two sexes can evolve adaptation and counter-adaptation, but it is not possible to make any more specific statements about the outcome of sexual conflict races.

The significance of courtship

As we have mentioned earlier in the chapter, some aspects of courtship behaviour can be interpreted in terms of sexual conflict and sexual selection. However this is not true of all courtship signals: many are designed for species identification, and here the interests of the two sexes are similar because both benefit by mating with a member of the same species. Some of the clearest examples of this role of courtship come from studies of frog calls. When several species of frogs live in the same pond, each has a characteristic and distinct mating call given by the male, and females are attracted only to calls of their own species. In some frogs (e.g. the cricket frog *Acris crepitans*) it has been shown that the females' selectivity of response results from the fact that the auditory system is tuned to the particular frequencies in the male call (Capranica *et al.* 1973).

Courtship displays may also play a role in competition between males within a species for mating opportunities. Often the same displays simultaneously serve to repel other males and attract females. An example for which this has been demonstrated experimentally is the mating call of the pacific tree frog (*Hyla regilla*) (Whitney & Krebs 1975a, b). Males are repelled and females attracted by loudspeakers broadcasting the mating call and females select out of a group of loudspeakers the one which calls for the longest bouts. Females may choose between displays purely on the basis of sexual attractiveness, as explained by Fisher's theory of sexual selection, but there is also the possibility that differences in courtship between males may indicate habitat quality, for example males with territories containing a lot of food might be able to afford to spend more time displaying.

A third role of courtship to which we have referred is assessment. In a species with male parental care females may assess the ability of the male to look after young and males may assess whether a female has previously been fertilised. Early work by ethologists on birds and fish showed that at the beginning of courtship males are often aggressive and females are coy or reluctant. Courtship was seen, therefore, as serving to synchronise sexual arousal of the partners. A possible explanation of *why* it should be necessary to overcome aggression and reluctance is that the early phases involve assessment by both partners before investing in offspring.

Throughout this chapter we have emphasised the role of females as investors in the zygote and offspring, but we have also mentioned that sometimes males invest as much as or more than females. Why should this happen in some species but not in others? In order to answer this question we will turn in the next chapter to the influence of ecological pressures.

Summary

Conflict lies at the heart of sexual reproduction. The fundamental difference between male and female is the size of gametes. Males produce tiny gametes and can be viewed as successful parasites of large female gametes. Because sperm is cheap and plentiful, males can increase their reproductive success by mating with many females. Females can only increase their success by making eggs or young at a faster rate. Females are a scarce resource for which males compete and much of male courtship can be understood in terms of competition for matings. Females may be reluctant to mate unless they can choose partners with good resources or (perhaps) good genes. Sometimes the general rule of high female investment is reversed and males are the main investors.

Further reading

Trivers (1972) is a seminal paper in which the relationship between parental investment and sexual selection was first spelled out.

Maynard Smith (1978b) reviews various aspects of sex including the sex ratio and anisogamy (a shorter version of a similar review is in Maynard Smith 1978c). Halliday (1978) reviews sexual selection and mate choice. Blum & Blum (1979) contains a number of articles of interest, for example Parker on sexual conflict. Lloyd (1979) has written a marvellously entertaining review of intra-male conflict and sexual selection in insects. O'Donald (1980) gives a population geneticist's view of sexual selection. The first two chapters and the critique of Zahavi's handicap principle are particularly relevant.

Williams (1979) is a thought-provoking discussion of the sex ratio. He points out that in vertebrates many adaptive modifications of the sex ratio which are predicted by extensions of Fisher's theory do not seem to occur. He raises the question of whether there is sufficient genetic variance for the sex ratio to evolve and whether an evolutionary parent-embryo conflict may have resulted in the inability of mothers to manipulate the sex of their offspring.

A particularly nice study of sexual conflict is that of Smith (1979). He shows how males of a species of bug ensure that they fertilise the eggs that they subsequently carry on their back.

Chapter 7. Parental Care and Mating Systems

In the last chapter we explored the fundamental differences between males and females. A male has the potential to father offspring at a faster rate than a female can produce them. Therefore females are a limiting resource for male reproductive success and males are expected to compete to maximise their number of matings. On the other hand, because females put a lot of investment into the zygote we expect them to resist male ardour and to be selective in their choice of mate.

However, care for the zygote does not end with gamete investment. In many animals there is some form of parental care, for example guarding the eggs or feeding the young. Parental care may be done by both parents (e.g. starling, *Sturnus vulgaris*), by the female alone (e.g. red deer) or by the male alone (e.g. sea horse, *Hippocampus*, where the male carries the fertilised eggs around in his brood pouch).

Often, differences between species in parental care are associated with differences in mating system. Mating systems can be categorised under the following broad headings.

1 Monogamy. A male and a female form a pair bond, either short or long term (part of or a whole breeding season or even a lifetime). Often both parents care for the eggs and young.

2 Polygyny. A male mates with several females, while females each mate with only one male. A male may associate with several females at once (simultaneous polygyny) or in succession (successive polygyny). With polygyny it is usually the female that provides the parental care.

3 Polyandry. This is exactly the reverse of polygyny. A female either associates with several males at once (simultaneous polyandry) or in succession (successive polyandry). In this case, it is the male who does most parental care.

4 Promiscuity. Both male and female mate several times with different individuals, so there is a mixture of polygyny and polyandry. Either sex may care for the eggs or young. *Polygamy* is often used as a general term for when an individual of either sex has more than one mate.

These divisions are not hard and fast but they will help as a general guideline. Because the mating system is often correlated with the kind of parental care, we could ask whether the mating system is a *cause* or a *consequence* of differences between the sexes in parental care. For example, in a monogamous bird like the starling the male does an appreciable amount of care. Is it right to say starlings are monogamous because males help care for the eggs and young, or male starlings have to help and therefore starlings tend to be monogamous? There is no straightforward answer to this question and so rather than attempt to discover which came first, it is more constructive to try and find out why different species have different types of parental care and mating system.

The ideal life for a male may be to go round copulating with lots of females, each of which stays at home to care for his offspring. For a female, the ideal may be to mate and then leave the male to care for the offspring while she forms reserves for more eggs. In practise, as we shall now see, there are two factors that will influence how the sexes resolve this conflict. First, different groups of animals have different physiological and life history constraints which may predispose one sex to do more parental care than the other. Second, ecological factors will influence the costs and benefits associated with parental care and mating behaviour.

Proximate constraints on parental care

These can be illustrated by contrasting three classes of vertebrates, fish, birds and mammals (Table 7.1). Are there any basic differences in the physiology and life histories of these three groups which could account for the differences in parental care and mating system? While considering this question we must bear in mind that both male and female will be selected to maximise their reproductive success and may behave at the expense of the other sex (Trivers 1972).

Table 7.1 Frequent types of parental care and mating system in three classes of vertebrates.*

	Parental care	Mating System
Birds	Both male and female	Monogamous
Mammals	Female only	Polygynous
Fish	Male only	Polygamy/Promiscuity

Note
*These are very broad generalisations and many exceptions are discussed later in the chapter.

BIRDS

As we saw in the first chapter, reproductive success in birds can be limited by the rate at which food is delivered to the nest. At least in those species where parental feeding of the young is important, we could imagine that two parents will be able to feed twice as many young as a single parent. Therefore both male and female increase their reproductive success by staying together. If either sex deserted, they would approximately halve the output of the last brood and would also have to spend time searching for a new mate and nest site before being able to start again. So monogamy and parental care by both male and female are not hard to understand. In some seabirds like the kittiwake, *Rissa tridactyla* (Coulson 1966) and the manx shearwater. *Puffinus puffinus* (Brooke 1978), long-term mate fidelity seems to pay because pairs that have bred together previously have higher reproductive success than new pairs.

When the constraint of having to have two parents feeding the young is lifted, it is usually the male who deserts and the female who is left caring for the brood. Comparative evidence shows that polygyny often occurs in fruit and seed eaters, probably because these food supplies become so seasonally abundant that one parent can feed the young almost as efficiently as two (e.g. weaver birds, chapter 2). Why is it the male who deserts? There are two factors which may be important. First, the male has the opportunity to desert before the female. With internal fertilisation, she is left literally holding the babies inside her. Second, the male can gain more by desertion than the female because his lifetime reproductive success depends more on his number of matings (see chapter 6). Another argument sometimes put forward to explain why the male deserts in species with internal fertilisation is that he can never be completely certain of his paternity and so has less definite genetic representation in the offspring then the female. However this may not influence a male's decision to desert because presumably he will be equally uncertain of his paternity if he deserts and mates again. A male will only gain by desertion if this increases his reproductive output above that which he would achieve by staying.

MAMMALS

In mammals, females are even more predisposed to care for the young. The offspring often have a prolonged period of gestation inside the female, during which the male can do little direct care (though he can protect and feed the female). Once the young are born they are fed on milk and only the female lactates. Because of these constraints on the opportunity to care for offspring and also because, with internal fertilisation, the male can desert first, it is not surprising that most mammals have a polygynous mating system and parental care by the female alone.

Monogamy and biparental care occurs in a few species where the male contributes to feeding (carnivores) or to carrying the young (e.g. marmosets). Perhaps it is surprising that male lactation has not evolved in these cases (see Daly 1979).

FISH

In the bony fish (teleosts), most families have no parental care (191 out of 245 families studied). In those families which do care for the eggs or young, it is usually done by one parent (46 out of 54 families). Compared with the elaborate care of offspring by birds, parental care in fish is a simple affair often consisting of just guarding or fanning the eggs. These tasks can usually be done effectively by one parent alone. If one parent can desert without affecting the success of the young which sex will it be? Table 7.2 shows that in families with internal fertilisation female care is commoner while with external fertilisation male parental care is most common.

Table 7.2 The number of families of bony fish in which, for some or all species, parental care is by the male, female or both sexes (after Maynard Smith 1978c; data from Breder & Rosen 1966).

		Parental Care	
	By male	By female	By both
External fertilisation	28	6	8
Internal fertilisation	2	10	0

The reason for this difference can be understood in terms of which parent has the opportunity to desert first and thus leave the other in a 'cruel bind'. As with the birds and mammals, internal fertilisation gives the male the chance to desert first. With external fertilisation, however, the roles are reversed. The female deposits the eggs first of all and then the male lays down sperm. Because sperm is lighter than eggs, the male must wait until the eggs are laid before he can fertilise them or else his gametes will float away. Therefore the female has the opportunity to desert first and swim off while the male is still fertilising her eggs (Dawkins & Carlisle 1976). Other factors may also predispose the male to care for the eggs when fertilisation is external. For example, he may have invested in a nest to which he can attract further females for mating (see next section). In some cases of external fertilisation, where male and female spawn close together without interference, the male may be more certain of his paternity than with internal fertilisation. However, paternity certainty should not bias a male's decision to stay with the fertilised eggs (see above).

ANCESTRAL ORIGIN OF UNIPARENTAL CARE

Another factor that may influence which sex cares for the zygote is the ancestral origin of uniparental care. In birds, the ancestral state is probably biparental care because in most species both male and female look after the eggs and young. This means that the female cannot use up all her resources in egg laying since she has to care for the young later on. It is well known that female birds can often readily lay replacement eggs if their first clutches fail. Therefore, if the male should desert, the female will still have some resources left to care for the brood.

In fish, the ancestral state is probably no parental care. The female may therefore use up all her resources in egg laying and so cannot afford to guard. This may mean that it is the male who often has to look after the young if care is necessary (Maynard Smith 1977a).

We have seen how differences between animals in parental care and mating system may be due to differences in proximate constraints such as the sequence of gamete deposition or physiological specialisations like female lactation. However the ultimate reasons for the differences must relate to ecological factors. For example, we are still left asking the question why do some fish have parental care and others not? In the next two sections we will explore how a species' ecology influences its mating and parental behaviour.

Ecological correlates of parental care in fish

The teleost fish are a good taxonomic group with which to illustrate that there are ecological correlates of parental care. They live in a wide variety of habitats and have a variety of types of parental care.

A comparative analysis of all the families for which there are good data shows that marine fish are much less likely to have parental care than are freshwater fish (Table 7.3). Baylis (in prep.) has suggested how ecological differences between the two habitats could account for differences in the frequency of parental care.

Table 7.3 The number of families of freshwater and marine fish which guard or do not guard their eggs and young (Baylis in prep.).

	Non-guarder	Guard	Per cent guard
Freshwater	23	31	57.4
Marine	104	19	15.4

The open ocean is one of the largest and most uniform environments on earth. It is like a giant homogeneous womb whose contents are constantly mixing. Any zygote released into the sea is likely to remain in the same thermal and chemical regime throughout its development. In contrast freshwater habitats, such as lakes and streams, are heterogeneous in space and time. Because they are smaller bodies of water there are greater effects of the substrate and air on water temperature, movement and chemical composition. In a temperate lake, for example, two points on the water surface only a hundred metres apart may differ by 10°C and a depth change of just 40 cm could mean a change in temperature of 4°C. If zygotes were left free to drift they would encounter changing conditions which could cause developmental abnormalities. Furthermore, it is likely that in freshwater bodies some micro-habitats will be better for zygote development and safer from predators.

Therefore in freshwater there will be much greater potential for fish to increase the success of their zygotes by choice of particular spawning areas. If good sites are scarce relative to the breeding population then there may be competition for them. Any male that was able to defend a good spawning site as a territory would attract many females and enjoy a high reproductive success. For example, in pupfish (*Cyprinodon* spp.) females prefer to spawn on rocky substrates and males defending these sites do much better than those with territories in sandy areas (Kodric-Brown 1977).

Once male territoriality has evolved, with females laying their eggs in the territory, there is the potential for the evolution of parental care by the male. He may gain by modifying the habitat to make it even better as a nursery for the eggs, for example by building a nest (e.g. stickleback). He could further improve zygote survival by guarding them against predators, fanning them to increase their oxygen supply, or removing any diseased eggs. When good spawn sites are localised and reusable it will also pay the male to stay because he can attract further females for laying. In the garibaldi (*Hypsypops rubicunda*) the difference in reproductive success between successful and unsuccessful males is enormous; males with nests average 129 000 eggs per nest while males without a nest do not breed at all (Clarke 1970).

The strategy of guarding eggs at a spawn site would not be suitable in very unstable habitats like shallow temporary ponds, tidal waters and estuaries. Where zygote development could be adversely affected by rapid changes in physical and chemical properties of the habitat, 'bearing' has evolved. This consists of keeping the eggs or young in the adult's mouth (e.g. Cichlidae) or in a brood pouch (e.g. sea horses, Syngnathidae) to protect them from environmental fluctuations. Bearing will be favoured as a form of parental care whenever the temporal variations in the habitat occur over a time shorter than that needed to produce viable larvae.

In summary, Baylis's survey shows that parental care in fish is correlated with spatial and temporal unpredictability of the environment for zygotes and young. With intermediate levels of unpredictability, care evolves via defence of good spawning sites and when the environment is extremely unpredictable brooding may evolve.

Ecological correlates of mating systems

We will now examine in detail how particular environmental factors influence male and female reproductive success and so determine the form of parental care and mating system. The ideas can best be illustrated with reference to different kinds of polygamy in birds and mammals. The male can usually benefit more than the female can by maximising his number of matings and so the major ecological influence on mating systems will be how environmental factors determine the ability of a male to gain access to females.

A male could increase his reproductive success either by defending females themselves or by controlling some critical resource that the females need and then mating with them when they visit the resource (e.g. spawning sites in the fish described above and the frogs in chapter 6). As we saw in chapter 4, the ability of an individual to control access to limiting resources will depend on the economics of defence. If resources or females are easily defendable then there will be high potential for polygyny. The economics of female or resource defence will be determined by their distribution in space and time (Emlen & Oring 1977; Wittenberger 1980).

(a) Spatial distribution

When mates or resources are evenly distributed there is little opportunity for one individual to command more than others. However, when they are patchily distributed some individuals may be able to defend more mates or better quality resources (e.g. nest sites, food) than others. A patchy environment can therefore be said to have a high polygamy potential (Fig. 7.1).

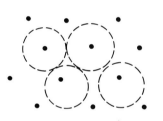
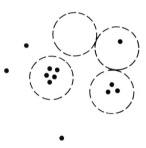

Even distribution.
Little polygamy potential

Patchy distribution.
High polygamy potential

Fig. 7.1 The influence of the spatial distribution of resources (food, nest sites) or mates on the ability of individuals to monopolise more than others. Dots are resources and circles are defended areas. With a patchy distribution of resources or mates there is greater potential for some individuals to 'grab more than their fair share'.

(b) Temporal distribution

The key factor here is the 'operational sex ratio' (see also chapter 6), which is the ratio of the number of receptive females to the number of sexually active males at any one time. If all the females were receptive at exactly the same time then, with a real sex ratio of 1:1 in the population, the operational sex ratio at the time of breeding would also be 1:1. There would be little opportunity for a male to mate with more than one female because by the time he had mated once, all the other females would have finished breeding. With female synchrony, therefore, there is little potential for polygyny. Nancy Knowlton (1979) has suggested that female synchrony may evolve as a mechanism to enforce monogamy on males. If, on the other hand, the females became receptive in sequence there would be a greater opportunity for a male to mate more than once. It is clear that the greater the operational sex ratio deviates from unity, the greater the potential for polygamy.

The influence of female synchrony on male mating success can be illustrated by comparing two species of anuran amphians. The common toad, *Bufo bufo*, is an explosive breeder; all the females lay their eggs in a period of about a week. This means that a male only has time to mate with one female, or at best two, before the breeding season has ended. Bullfrogs, *Rana catesbeiana*, in contrast, have a prolonged breeding season with females arriving at the pond over several weeks. Males which can defend the most attractive spawning sites may mate with as many as six females in a season.

We shall now examine how the evolution of polygamy is determined by the spatial and temporal distribution of resources and females. To make the discussion clearer we will divide polygamy into various ecological categories.

RESOURCE DEFENCE POLYGYNY

There are many examples where the males that defend the best quality food patches or nest sites gain access to several females while males with poor resources never mate. Male orange-rumped honeyguides, *Indicator xanthonotus*, defend bees' nests and when a female comes to feed on the wax the male copulates with her. In effect the male is trading food for sex; the more bees' nests he can defend, the more females he will attract. One territorial male was seen to mate with 18 females while non-territorial males failed to mate at all (Cronin & Sherman 1977).

The problem for a female who mates with an already paired male is that she may suffer a decrease in reproductive success because the resources or care by the male have to be shared with other females. In the yellow-bellied marmot (*Marmota flaviventris*), males who defend the best burrow sites attract the most females. However, the more females that breed in the same burrow system, the lower the success per female (Fig. 7.2). The best mating system from the female's point of view appears to be monogamy. However the male has most success with two or three females (Fig. 7.2). The number of females observed on a territory is usually about two, so this may represent some sort of a compromise between the conflicting interests of each sex.

Jared Verner and Mary Willson (1966) suggested that polygyny could arise by female choice for good resources (e.g. nest sites, food) if the difference in quality between male territories was great enough to mean that a female could rear more young by sharing a good territory with another female rather than by mating monogamously with a male on a poor territory. Gordon Orians (1969) summarised this idea in a

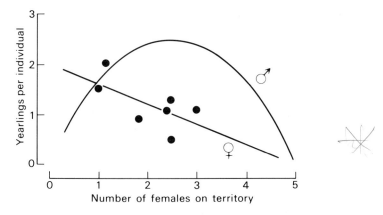

Fig. 7.2 Reproductive success in yellow-bellied marmots in relation to the number of females on a male's territory. The success per female declines with number of females (solid dots). The success of the male, which is simply the success per female multiplied by the number of females, peaks at 2–3 females (Downhower & Armitage 1971).

graphical model which is explained in Fig. 7.3. The model makes the following predictions. First, if it is differences in the quality of male territories that influences the evolution of polygyny then we would predict that polygyny would occur most often in 'patchy' habitats where differences in territory quality will be most marked (Fig. 7.1). Verner & Willson (1966) tested this prediction by comparing the habitats and mating systems of all the North American passerine birds. Of the 291 species, 14 were regularly polygynous or promiscuous. Thirteen of these bred in marshes, prairies or savannah-like habitats where there are often big differences in productivity even between small adjacent areas. There are two problems in the interpretation of these data. One, that we have encountered before, is the danger of invoking 'patchy resources' as an explanation without rigorous measurements of resource distribution and abundance. The other is the problem of which is the appropriate taxonomic level for analysis of this comparative data. Nine of the 14 polygynous species are in one family, the Icteridae, and it is difficult to know whether we should treat these as nine independent cases or one (see chapter 2 for a discussion of this problem). Nevertheless, Verner and Willson's analysis does suggest that differences in male territory quality may be an important ecological correlate of polygyny.

The second prediction from the model is that a male's territory quality will influence his reproductive success, in particular the number of females he attracts. The two most important resources that will influence success are probably food availability and nest sites. In the long-billed marsh wren (*Telmatodytes palustris*) males with the largest territories, which contain the most food, attract two females while those with small territories attract only one mate or remain as bachelors (Verner 1964). In red-winged blackbirds (*Agelaius phoeniceus*) males whose territories contain the type of vegetation that provides the safest nest sites are the ones that attract the most females (Holm 1973).

One of the clearest examples of how a male's territory quality influences female reproductive success is Wanda Pleszczynska's (1978) study of the lark bunting (*Calamospiza melanocorys*). This bird is a summer visitor to the grasslands of North America. The males arrive first in the spring, compete for territories and then display with song flights to attract females which arrive soon afterwards. Once a female settles on a territory she builds a nest and the male helps her care for the offspring. Pleszczynska found that some males attracted a second female even when there were unmated males available in the population. These 'secondary females' got no help with breeding from the male, who was already fully occupied with his first mate, and their breeding success was lower. Nevertheless they still chose to mate polygynously. The resource on a territory which was most important for reproductive success was a shady nest site. The main cause of nestling mortality was over-heating in the hot sun and observations showed that nests built in the shade had much greater fledgling success. This was also demonstrated experimentally by putting up small strips of plastic above some exposed nests to act as umbrellas and so protect the young from the sun. The success of these experimental nests was significantly greater than that of unshaded controls. As we might predict, the males who attracted two females were those which defended territories with the most cover. Pleszczynska was even able to go to a new area and predict the mating status of a male (bigamous, monogamous, bachelor) before the females arrived simply from a shading score for each territory.

The third prediction from the model in Fig. 7.3 is that, for females settling at the same time, the reproductive success of those who choose to mate with an already-paired male will be at least as great as those who choose an unpaired male. For example, in the lark bunting a female who mates polygynously is expected to make this choice because access to a better nest site offsets the disadvantage of having to rear the young without any male help. As predicted, Pleszczynska found that secondary females had fledgling success at least equal to that of monogamous pairs breeding at the same time. It is worth pointing out the difficulties of measuring reproductive success to test this prediction. As we saw in the great tit study, described in chapter 1, the number of young fledgling from a nest may not always give a good indication of success because fledgling weight will influence survival. Furthermore, a female is presumably selected to maximise her lifetime reproductive success which may not be the same as maximising the number of young produced each year because her chances of surviving to breed again decreases with the brood size she produces.

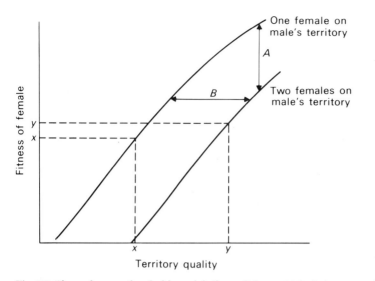

Fig. 7.3 The polygyny threshold model (from Orians 1969). It is assumed that reproductive success of the female is correlated with environmental factors, such as the quality of the male's territory in which she breeds, and that females choose mates from the available males. In the model, a female suffers a decrease in fitness A by going to an already mated male compared with the fitness she could expect if she had a male all to herself. Despite this drop in fitness, provided the difference in quality between the territories is sufficient (B=the polygyny threshold), a female may expect greater reproductive success by breeding with an already mated male. For example, a female who shared a male on territory y would do better than a female who had a male all to herself on territory x.

FEMALE DEFENCE POLYGYNY: HAREMS

Polygyny can also occur without resource defence when males defend the females as a harem. Females will be most economically defendable when they are in groups, for example to decrease predation or increase feeding efficiency (see chapter 4). Polygyny can be regarded, therefore, as a consequence of selective pressures favouring grouping in females. Harem defence is particularly common in mammals.

Female Northern elephant seals, *Mirounga angustirostris*, haul up on beaches to drop their pups and mate again for the production of the next year's offspring. Because the females are grouped in a few localised areas they are a defendable resource and the males fight with each other to monopolise them. The largest, and strongest, males win the biggest harems and in any one year almost all the matings are performed by just a few males (Le Boeuf 1972, 1974). The Southern elephant seal, *M. leonina*, has a similar mating system (McCann 1981; see Fig. 7.4).

To be a harem master is so exhausting that a male usually only manages to be top ranking for a year or two before he dies. He has to fight off other males continuously and in the process of defending his harem he sometimes tramples on his females' new-born pups. Although this is obviously not in the females' interests, these pups were

(a)

(b)

(c)

Fig. 7.4 (a) (see facing page) Female Southern Elephant seals haul up on beaches to give birth. (b) (see facing page) The large males fight to defend harems. (c) Successful males copulate with the females (Photos by T. S. McCann).

probably not sired by the male himself because he is unlikely to have been a harem master the year before. From the male's point of view, therefore, there is little cost in damaging or even killing the pups; his main concern is to protect his paternity for next year's offspring (Le Boeuf 1974; Cox & Le Boeuf 1977).

MALE DOMINANCE POLYGYNY: LEKS

Finally, there is sometimes polygyny even without resource defence or female defence. Instead, mating priority is determined by the dominance status of the male. In some birds and mammals the males congregate on communal display grounds called leks. They display to one another and defend tiny territories. In manakins, *Manacus m. trinitatis*, these are just bare patches of ground no more than a metre or two across and they do not contain any obvious vital resources (Lill 1974). Females visit the leks and most of them choose to mate with one of the males in the centre of the group. One or two of the males on each lek end up performing most of the copulations (Fig. 7.5). Male hammer-headed bats, *Hypsignathus monstrosus*, also congregate at traditional leks where over a hundred may set up small territories in the tree canopy and call to attract females. These assemblies do not contain food resources and are simply sites where the females come to mate. Once again most of the copulations are performed by just a few of the males. For example in one lek of 85 males, 79 per cent of the copulations were accounted for by 5 males (Bradbury 1977).

The precise ecological conditions favouring this kind of polygyny have not been worked out. Lek mating systems may have evolved where neither the females themselves nor the resources they require are economically defendable and where the females are able to care for the offspring alone. The males may then congregate together to display because this decreases predation, or female choice may have favoured

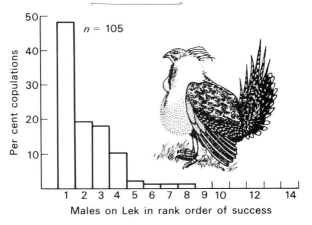

Fig. 7.5 In sage grouse (*Centrocercus urophasianus*) leks almost all of the copulations are performed by just a few of the males. The most successful males are those that defend the central territories on the lek (from Wiley 1973).

communal displays because they can then choose the 'best' male more easily. However we still have little idea as to what factors make a male the 'best' when his sole contribution to the offspring is sperm (chapter 6). It may simply be that the centre of the lek, where females prefer to mate, is the safest place from predators.

In summary, polygyny may evolve when males defend good resources, or when they are able to defend groups of females or when they compete directly for access to females in a lek.

POLYANDRY IN BIRDS

Polyandry is much rarer than polygyny because it is usually the males who will gain the most by maximising their number of matings. Nevertheless there are a few species of birds where the behaviour of the sexes is reversed and females compete with each other for access to males (Jenni 1974).

The evolutionary pathway to polyandry is most easy to understand in the Arctic waders (Oring 1981; Maxson & Oring 1980). In some species, like the Temminck's stint, *Calidris temminckii*, and the sanderling, *Calidris alba*, the male defends a territory. The female lays one clutch of eggs and the male incubates them while she lays another clutch and then incubates that herself. Therefore the pair produces two broods, one cared for by the male and one by the female. This is an efficient strategy for breeding in the arctic tundra where the season is very short but very productive. The female is able to form two clutches in rapid succession because the insect food supply is so rich, and the neat division of labour between the male and female enables them to raise two broods during the short period of high productivity. Two clutches are also a better insurance against predation, which can be heavy.

This system of 'double clutching' could give rise to the evolution of polyandry if the female is not faithful to the first male but, instead, goes off to lay her second clutch for another male. As food abundance increases so that a female can produce many clutches then she will do best by laying for a number of males in turn.

This is exactly what has happened in the spotted sandpiper, *Actitis macularia*, where productivity on the breeding grounds can be so high that the female becomes rather like an egg factory (Fig. 7.6). A female can lay 5 clutches in 40 days, a total of 20 eggs, which represents 4 times her own body weight! Her reproductive success is no longer limited by her ability to form reserves for the eggs but rather by the number of males she can find to incubate them. At some times most of the males are tied up with incubation duties. These ecological conditions have given rise to the evolution of sex role reversal. The females are 25 per cent larger than the males and they compete with each other to get males to care for their clutches. Females defend territories and some may have three males sitting on clutches at the same time. Sometimes the female may incubate the last clutch of the year herself.

Fig. 7.6 A polyandrous wader, the spotted sandpiper. This photograph by Lew Oring shows two brood males in an agonistic encounter at a territory boundary. Female aggression tends to be more intense and peaks at the start of the season when females are competing to acquire males.

Ecology and dispersal

As well as influencing the evolution of parental care and mating systems, ecological factors are also an important determinant of another aspect of reproduction, namely dispersal. The movement of a young animal from its place of birth to that of its first breeding attempt is referred to as natal dispersal. When birds and mammals are compared, some striking trends emerge (Greenwood 1980; Table 7.4).

Table 7.4 The number of species of birds and mammals where natal dispersal is more extensive in males or females (from Greenwood 1980).

| | No. species with predominant dispersal | | |
	By males	By females	No sex difference
Birds	3	21	6
Mammals	45	5	15

1 *In both birds and mammals, one sex usually disperses more than the other.*

The result is that close inbreeding is avoided. When an animal mates with a close relative there is a greater chance that any harmful recessive alleles will become homozygous in the offspring and cause low reproductive success.

For example, in the great tit population in Wytham Woods, Oxford, juvenile females disperse more than males so that from a total of 885 pairings only 13 (1.5 per cent) have been inbred pairs. These rare cases were where a male had dispersed more than average or where a female had dispersed less, with the result that a son mated with his mother or a brother with a sister. Such inbred pairs had lower reproductive success than outbred pairs (Greenwood, Harvey & Perrins 1978). It is known that inbreeding also results in decreased reproductive success in mammals and it is one of the problems associated with trying to breed from small populations in zoos (Ralls, Brugger & Ballou 1979).

Another way of avoiding inbreeding would be to recognise close relatives. This could come about through early learning because the individuals most closely encountered when young are likely to be kin. This mechanism may operate in man; in a study of Israeli kibbutzim, where children are brought up communally, it was found that individuals never married anyone with whom they were associated when young, even when they were not close relatives (Shepher 1971; see also chapter 13).

2 *In birds, females disperse more than males.*

As we saw above, in many species of birds the male helps care for the young. Often he defends a territory and females may choose a male on the basis of his territory quality. It may pay a male to remain near his birth site because it might be easier to set up a territory in the vicinity of relatives, for example by inheriting part of the father's territory (see chapter 9). Once this happens, it may pay the females to disperse to avoid inbreeding. They may also benefit by moving so that they are able to sample many male territories and choose the best.

3 *In mammals, males disperse more than females.*

Male mammals are more often polygynous than birds, and their mating system is usually based on mate defence rather than resource defence, with the male contributing little to the care of the offspring. In mammals, a male will benefit most from gaining access to a large number of females and so male dispersal may have been favoured.

It is clear that much more data are needed on the costs and benefits of dispersing to breed versus staying at home. For any one individual the pay-offs of the two options must depend on what others in the population are doing and so we really need a theory which will analyse the problem in terms of what will be the evolutionary stable strategy for a male and a female. We should also remember that in many cases, particularly mammals where the male contributes little to parental care, the cost of inbreeding for a male will be less than that for a female. For example in the olive baboon (*Papio anubis*) inbreeding is avoided because males transfer from their natal troops whereas females do not. However, before they disperse, young males may still surreptitiously try to mate with females, even close relatives. The females combat this risk of incestuous mating by preferring to associate with transferred (strange) males rather than natal (familiar) males who are likely to be relatives (Packer 1979).

CONSEQUENCES OF SEX DIFFERENCES IN DISPERSAL

A consequence of differential dispersal of the sexes is that members of the sedentary sex will tend to be closely related to their near neighbours and so we may expect more altruism between them than between members of the dispersing sex where individuals living near by will tend to be unrelated.

In mammals, where females are usually more sedentary, the females in an area are often relatives. Altruism between females is common; for example female ground squirrels give alarm calls to warn other females of the approach of a predator (Sherman 1977) and in lionesses, females will suckle other females' cubs (Bertram 1975).

In birds it is the other way round. Males disperse less and so in a particular area it is the males who are closely related and who show most altruism to each other. In chapter 9 we will encounter several species of birds where the males help look after another individual's offspring.

Conclusion

It is clear from our comparative approach to different groups of animals and different habitats that ecological factors are important in shaping parental care and mating system. The distribution of resources, such as suitable egg-laying sites and food, and the distribution of the females in space and time all influence the way individuals can behave to maximise their reproductive success. Proximate constraints, like the mode of fertilisation, may also determine the mating system by the way they predispose one parent to care for the offspring and thus give the other the opportunity to desert and go off in search for further mates. We have also seen that the evolution of the behaviour we now see will depend on what came beforehand. Which sex guards the offspring in uniparental care in fish and birds, and the evolution of polyandry in waders are both easier to understand if we take account of the ancestral state of the behaviour.

Categorising species in terms of their characteristic 'mating system' is fine when we are making broad comparisons across taxa or across habitat types. However we should not expect all individuals within a species to behave in exactly the same way. While some individuals are displaying or defending a resource to attract mates, others may be achieving reproductive success by more devious and sneak methods. In the next chapter we will take a closer look at the evolution of these alternative strategies.

Summary

Differences between species in parental care and mating systems can be correlated with differences in physiological constraints and ecology. Internal fertilisation and specialisations such as lactation predispose females to perform parental care, while external fertilisation and the need for two parents to look after the young may constrain males to show parental care. Parental care in fish, by either sex, tends to be correlated with environmental unpredictability. In birds and mammals polygyny occurs when males defend good resources (e.g. lark bunting) or groups of females (e.g. elephant seals). There are sex differences in dispersal in birds and mammals which might also be related to ecological factors.

Further reading

An important paper on parental care and mating systems is that by Maynard Smith (1977a) who uses an ESS approach, emphasising that the best strategy for one parent must depend on what the other parent is doing. Ridley (1978) reviews paternal care. Good reviews of mating systems in particular animal groups are those by Bradbury & Vehrencamp (1977) on bats, Wells (1977) on frogs and toads, Pitelka, Holmes & MacLean (1974) on arctic wading birds, and Jouventin & Cornet (1980) on seals. Alexander (1975) discusses chorusing behaviour and leks in insects. The paper by May (1979) examines the costs and benefits of staying at home and being incestuous versus dispersal and outbreeding.

Chapter 8. Alternative Strategies

In the last chapter we used mainly a comparative approach to try and understand differences between species and we characterised each by a 'typical' mating system. During recent years, however, it has become clear that there are marked differences between individuals within a species in the way they compete for scarce resources. A decade ago, if an animal was seen behaving in a diferent way from the majority of the population it was often thought to be abnormal. Male ducks that raped females instead of courting them by displays were said to be behaving abnormally due to overcrowding. If we observed a male bullfrog sitting silently in the middle of a chorus, while other males were croaking loudly to attract females, we would, perhaps, have thought that it was ill or having a rest.

Nowadays, whenever we see an individual doing something different we are tempted to label it as a 'strategy'. Silent male frogs may not be tired after all, they may be employing sneaky strategies. There are three main reasons for this change of emphasis. First, the realisation that evolutionary arguments must be framed in terms of benefit to the individual or gene, rather than for the good of the species, has led us to expect individuals to compete selfishly with each other. If some males are attracting females by calling then we now expect to find others parasitising their efforts and behaving as sneaks. Once we have seen courtship as a conflict between individuals (chapter 6), rather than as a co-operative venture, we are not surprised to see some males attempting to rape females.

Second, the application of game theory to the study of behaviour (chapter 5) has shown that it is possible, in theory, to have stable equilibria with individuals in a population behaving in different ways. Finally, an increase in the number of field studies with individually marked animals has shown that indeed there are often several different strategies used within the same species to compete for a mate, a nest site or some other scarce resource.

In this chapter we will examine some examples of individual differences in competitive behaviour and discuss how they might have evolved.

The evolutionary pathways to alternative strategies

Three hypotheses can be suggested to explain why individuals within a population may behave in different ways.

CHANGING ENVIRONMENT

The best strategy may depend on the habitat, and if this is patchy in space or changes frequently in time then several strategies may persist. For example, the black form of the peppered moth is favoured in polluted habitats while the pale form does best in unpolluted areas. As long as both habitats persist, both colour morphs will survive (Kettlewell 1973). In some cases, individuals may change their behaviour depending on the habitat. When a male speckled wood butterfly searches for females in patches of sunlight on a woodland floor, it defends a territory and sallies out from favourite perches to inspect passing objects. When the same male flies up to the tree canopy, where females are scarcer, it adopts a different searching behaviour and patrols over a large area.

Another example, which probably falls under this heading, is the different colour morphs of the three-spined stickleback (McPhail 1969; Semler 1971; Moodie 1972). Some males have bright red throats while others have dull throats. The red males are more attractive to females; in a choice test with a red male at one end of a tank and a dull male at the other, most females selected the red male. The attractiveness of a dull male can be increased simply by dying his throat red, so it is the colour itself that is attractive to females rather than something special about the behaviour of red males. Why, then, are not all males red?

The answer is that although a red throat brings a benefit in terms of increased attractiveness to females, it also imposes a cost because red males are more susceptible to predation by trout. Experiments showed that red males were particularly preyed upon in bright light, when their colour enabled the predator to spot them more easily. Samples from North American lakes reveal that in deep water most males are of the red type. Presumably in these dark waters the mating advantage of red males outweighs the disadvantage of increased predation. In shallow waters, however, nearly all the males are dull in colour. Here, in the brightest light, dull males do best because the red males are very conspicuous and quickly eaten by predators.

The colour of sticklebacks is therefore selected for by a trade-off between mating and predation. Both red and dull males persist because each does better in a different environment.

MAKING THE BEST OF A BAD JOB

Sometimes, perhaps because of its small size, an individual is unable to compete successfully by fighting or displaying. Instead, it must make the best of its poor circumstances by employing some alternative strategy. Even though it ends up with fewer rewards than others, it will still be doing the best it can given its own competitive ability.

Body size is often age dependent. Unable to compete with larger rivals, young animals may attempt to steal resources by sneaking. We have already seen in chapter 6 that the largest bullfrogs win the best territories and attract females by croaking. Small, young males are not strong enough to defend a territory and so they behave as satellites, sitting silently near a calling male and attempting to intercept and mate with any females he attracts (Fig. 8.1). They are not very successful; only 2 out of 73 matings were by satellite males (Howard 1978). Nevertheless this is probably their best chance of getting a female when they are small, and the satellites make the best of a bad job by parasitising the largest calling males on the best territories, where most females will be attracted.

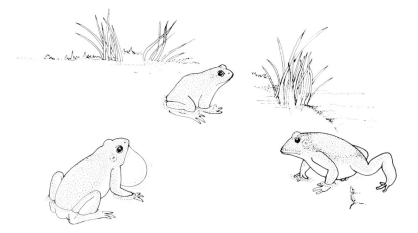

Fig. 8.1 A male bullfrog (left) calls from his territory and attracts a female who is entering the pond on the right. In the middle of the picture is a small male who sits silently in the larger male's territory as a satellite and attempts to intercept the female on her way to the caller.

Young male elephant seals have no hope of defending a harem themselves and instead they attempt to sneak copulations by pretending to be a female and joining a large bull's harem! (Le Boeuf 1974). Both young bullfrogs and young elephant seals will change their strategies from sneaking to fighting as they get older and grow larger and stronger.

In other cases, small body size may be fixed throughout an individual's lifetime and be a consequence of poor feeding conditions when young. An example is the bee, *Centris pallida*, where the largest males are three times the weight of the smallest males (Alcock, Jones & Buchmann 1977). Large males search for females by patrolling over the ground, searching for buried virgin females about to emerge. When they discover an emerging female they dig her up and copulate. It takes several minutes to dig up a female, during which time other males are attracted to the site by the activity. There are often violent fights and only large males are able to defend their discoveries successfully.

It is not surprising, therefore, that only large males adopt the strategy of patrolling and digging. Small males search for mates by hovering above the emergence areas and pursuing airborn females who have escaped the diggers. Intermediate sized males may adopt both digging and hovering. Observations showed that large males clearly had the greatest mating success and so it is probable that the smaller males are forced to adopt hovering throughout their lives to make the best of a bad job.

These are all examples of strategies that are conditional on an individual's phenotype, for example 'if big, fight; if small, sneak'. The largest and strongest individuals have the greatest success and others are forced by circumstance to adopt less successful, alternative strategies.

ALTERNATIVE STRATEGIES IN EVOLUTIONARY EQUILIBRIUM

Even when there are no constraints from the environment, or phenotypic constraints such as body size, individuals may still differ because the pay-off for one strategy depends on what others in the population are doing. We have seen examples of this in chapter 5 where different fighting strategies, such as Hawk and Dove, may coexist as a mixed ESS. The same argument applied to the evolution of the sex ratio, discussed in chapter 6, where the best sex to have as offspring depends on the ratio of males to females in the population; male and female can be regarded as different strategies, stabilised by frequency dependent selection as a mixed ESS.

If two strategies we observe in nature are an example of a mixed ESS then at equilibrium we would expect them to enjoy, on average, equal success, just like male and female or Hawk and Dove in our hypothetical game in chapter 5. This is a very different prediction from
the previous section where individuals employing a strategy that makes the best of a bad job are expected to have lower success than others.

Examples of alternative strategies

We will now look at some examples where individuals within a population differ in their behaviour and consider how these differences might have evolved, bearing in mind the three hypotheses discussed above. The strategies we will discuss are summarised in Table 8.1.

CALLERS AND SATELLITES

In the bullfrog, mentioned above, only the large males call to attract females, while males are satellites. Their strategy is therefore conditional on their phenotype, 'if large, call; if small, be a satellite'. In the treefrog, *Hyla cinerea*, however, there is no difference in body size between callers and satellites. It is unlikely, therefore, that satellites

Table 8.1 Examples of alternative strategies within the same species.

Species	Reference	Strategy 1	Strategy 2
Stickleback	Semler 1971	Red	Dull
Bullfrog	Howard 1978	Caller	Satellite
Treefrog	Perrill, Gerhardt & Daniel 1978	Caller	Satellite
Field cricket	Cade 1979	Caller	Satellite
Ruff	van Rhijn 1973	Resident	Satellite
Digger wasp	Brockmann, Grafen & Dawkins 1979	Digger	Enterer
Bee	Alcock, Jones & Buchmann 1977	Patroller	Hoverer
Figwasp	Hamilton 1979	Fighter	Disperser
Scorpionfly	Thornhill 1979	Hunter	Stealer

are simply young males who are making the best of a bad job because they do not have very fine voices. Could this be an example of a mixed ESS?

It is easy to see in theory how the benefits of caller and satellite could be frequency dependent. If all males were callers, many females would be attracted and satellite would be a very profitable strategy. If all males were satellites, on the other hand, then no females would arrive and so it would pay males to call. If frogs were free to choose between strategies (no body size constraints, for example) then we might expect frequency dependent selection to stabilise the frequency of callers and satellites so that they each enjoyed equal success.

Perrill, Gerhardt & Daniel (1978) measured the success of the two strategies by releasing females near callers and their satellites. The calling male amplexed with the female on 17 occasions while the satellite did so on 13 occasions. Therefore, unlike bullfrogs, these treefrog satellites achieved considerable success and the data are consistent with a mixed ESS interpretation, because the success of the two strategies in this experiment are not significantly different. However there is a difficulty in testing a theory which predicts equality in success; it is impossible to demonstrate statistically that two strategies have exactly equal pay-offs. All we can do is to infer equality if we fail to find a significant difference. The problem is that the smaller the sample size, the less likely we are to find a significant difference.

Another problem faced by the field worker in measuring the pay-offs of different strategies is that although it is often easy to measure benefits, such as number of females attracted, it is difficult to measure the costs. For example, even if we find that satellites attract fewer females, their net benefit may be just as high as that of callers because they may incur lower costs. Calling involves increased energy expenditure and also may increase predation. Howard (1979) found that some of his calling bullfrogs not only attracted females but also snapping turtles in search of an easy meal!

A good example of the cost of calling is William Cade's (1979) study of the field cricket, *Gryllus integer*. Territorial males call by rubbing their forewings together and, just as with the frogs, other males sit silently near the callers and attempt to intercept females as they arrive.

Cade was able to show, by broadcasting male calls from a loud-speaker, that calling not only brings benefits but also considerable costs (Table 8.2). Louder calls attracted more females but they also

Table 8.2 When the song of a male cricket, *Gryllus integer*, is broadcast from a loudspeaker, it attracts not only females but also satellite males and a parasitic fly which will kill the cricket (Cade 1979).

Broadcast	No. females	Attracted to speaker No. satellite males	No. parasitic flies
Silent	0	0	0
80 dB song	7	7	3
90 dB song	21	16	18

attracted more satellite males, who courted the females around the loudspeaker. The calls also attracted a parasitic tachinid fly (*Euphasiopteryx ochracea*) which laid larvae on the loudspeaker! When Cade sampled crickets he found that 11 out of 14 calling males were parasitised by larvae of this fly while only 4 out of 29 satellites were infected. Parasitism must inflict a serious cost on callers because the cricket always dies when the parasite's offspring emerge from its body (Fig. 8.2).

RUFF LEKS: RESIDENTS AND SATELLITES

Male ruffs gather on leks (see chapter 7) in the spring where they display to attract females for mating. The males are larger than the females and have a distinctive breeding plumage with elaborate neck ruffs and head tufts (Fig. 8.3). Some males are 'residents'; they defend little bare patches of ground about 30 cm in diameter and they chase off aggressively other residents who encroach on their territories.

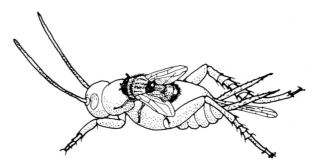

Fig. 8.2 Calling male crickets not only attract females but also a parasitic fly which lays live larvae into the cricket which will eventually kill it.

Fig. 8.3 Male ruff (*Philomachus pugnax*) display on leks to attract females. Females are smaller and lack the bright plumage and adornments of the males. The dark males on the left and right of the picture are territory residents. The white male in the foreground is a satellite.

Other males are 'satellites'; they are not aggressive and they associate with a resident on his territory. The resident sometimes tolerates the satellite and sometimes chases it away (Hogan-Warburg 1966; van Rhijn 1973).

The most striking observation is that this behavioural difference is correlated with a marked difference in plumage colour. Resident males have dark head tufts and usually dark ruffs as well. Satellite males have white tufts and ruffs. Males tend to have the same plumage throughout their lives and the difference between residents and satellites has a genetic basis.

When a female arrives on a territory she is courted by both the resident male and his satellite. Sometimes the resident chases the satellite away and then copulates with the female. However sometimes, while the resident is busy defending his territory against other males, the satellite copulates with the female. It is difficult to measure the reproductive success of the two strategies because the satellites may visit several leks and the female may copulate more than once. There also seem to be marked differences in the success of satellites at different times of the breeding season and on different leks. For example, on one small lek satellites were about equally successful as residents in their rate of copulation but on a larger lek the residents were very aggressve towards the satellites and they had almost no success at all. It is not yet clear whether the two strategies are an example of a mixed ESS or whether both strategies persist because each is advantageous in a different habitat or at a different time.

FIGWASPS: FIGHTERS AND DISPERSERS

Gadgil (1972) pointed out that within a population there could be selection for male dimorphism. Some males might put their resources into high competitive ability and have bright colours or weapons. This life style would be 'fast and furious', the male mating at a fast rate but living for only a short time because of the high costs involved. Other males might put their resources into survival and although they may reproduce at a slower rate this disadvantage may be offset by a longer lifespan.

A particularly dramatic example of male dimorphism within the same species occurs in some figwasps *Idarnes* spp. (Hamilton 1979). Some males are wingless and put their resources into fighting; they have large heads and mandibles that can chop another figwasp in half. These males remain inside the fig where they were born and fight to mate with newly hatched females that develop from larvae in the

Fighting male

Dispersing male

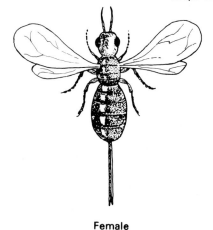

Female

Fig. 8.4 Male dimorphism in figwasps (from Hamilton 1979). Some males have small heads and can fly; others are flightless and have enormous mandibles that can chop another male in half.

fig. Other males are winged and put their resources into dispersal; they have tiny heads and mandibles, are not aggressive and fly off to mate with females that have emerged from the fruits (Fig. 8.4).

Although it is not yet certain how this dimorphism is maintained, it might well be a case in which the best strategy depends on the environment. For example, if figs contain few emerging females dispersal will pay. If, on the other hand, many individuals are born in the same fruit a male would have access to several females and it may pay to stay at home and fight.

NESTING STRATEGIES IN FEMALE DIGGER WASPS

Female golden digger wasps, *Sphex ichneumoneus*, lay their eggs in burrows underground which they provision with katydids as food for the offspring (Fig. 8.5). Jane Brockmann discovered that females have two ways of obtaining a burrow; they either dig one for themselves or they enter an already dug burrow. Digging a burrow is hard work and takes on average 100 minutes so entering may seem a good strategy because the female gets a burrow without having to spend time and energy in digging. However the wasps do not seem to recognise whether the burrow they enter is empty (abandoned by its builder) or occupied. If it is empty, then they are able to lay and provision in peace, but if there is another female using the same burrow the two wasps eventually meet and there is always a fight with the result that only one female is successful in breeding in the burrow. Even if a female digs a burrow herself she may be joined by another female who is playing the entering strategy (Brockmann & Dawkins 1979).

Fig. 8.5 A female digger wasp, *Sphex ichneumoneus* at a burrow entrance. Photo by Jane Brockmann.

How can we explain the evolution of these alternative strategies, 'digging' and 'entering'? It can be readily seen how their success could be frequency dependent. Imagine that all the females in the population are digging all the time. There would be plenty of empty burrows from previous and failed breeding attempts and so it would pay any female that started entering because she would save time not having to dig. Digging is therefore unlikely to be an ESS. However entering would not spread to takeover the whole population because if all the wasps were entering most would end up sharing, there would be lots of fights and it would certainly pay a female to go off and dig her own burrow, where the chances of sharing would be decreased. Entering is therefore also unlikely to be an ESS. However, because each strategy does best when rare (just like Hawk and Dove), there could be a stable mixture of entering and digging (a mixed ESS) brought about by frequency dependent selection, where the success of the two strategies was equal.

Brockmann, Grafen & Dawkins (1979) measured the success of the two strategies in terms of the number of eggs laid per unit time. Because individual females employed both strategies, they could not measure success by comparing *individuals* who were enterers with those who were diggers. Instead they measured the success of 'entering' and 'digging' *decisions*. Going back to the Hawk-Dove game in chapter 5, it will be remembered that a mixed ESS can come about either by a polymorphism in the population, some individuals always playing Hawk and others always playing Dove, or by individuals themselves playing Hawk and Dove in the proportion that satisfies the mixed ESS. Therefore, if the two strategies are an example of a mixed ESS we would predict that digging and entering decisions should have equal success.

To test this hypothesis data on the nesting behaviour of 68 female wasps were analysed. This field work was based on over 1500 h of observation and provided a nearly complete record of the histories of 410 burrows. Calculating the eggs laid per unit time from digging versus entering was a complicated exercise. Brockmann, Grafen & Dawkins (1979) had to work out the possible outcomes of the two decisions in order to calculate the overall success (Fig. 8.6). For example, if a female who was entering joined an occupied burrow and was then later evicted before she had the chance to lay any eggs, this had to be included as time spent 'entering' for no rewards.

The results showed that a female's decision to enter or dig was not conditional on any obvious phenotypic character, such as body size, nor on the environment, for example time of breeding season. Furthermore, there was no significant difference between the success rates of the two strategies. Therefore the hypothesis that best explains the data is that digging and entering are a mixed ESS. Individual females seem to be programmed with a simple rule such as 'Dig with probability p. Enter with probability $(1-p)$'. The value of p has

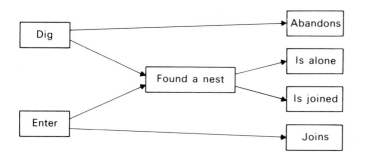

Fig. 8.6 A female digger wasp can either decide to dig a nest burrow herself or enter an already dug burrow. A digging decision may result in one of three outcomes; the wasp may abandon the burrow, she may start a nest and remain alone or she may be joined by another female. An entering decision may also result in one of three outcomes; a wasp may join another female in her nest, or she may have found a nest of her own in which case she may end up nesting alone or in company with another wasp. In order to calculate the overall benefits per unit time for the two decisions, the benefits and times spent in each outcome must be measured (from Brockmann, Grafen & Dawkins 1979).

evidently been fixed by frequency dependent selection so that the two decisions have the same reproductive success.

Problems of interpreting data

We have now described examples in which alternative strategies appear to be maintained in the three ways outlined at the beginning of the chapter. The figwasps and ruffs may be examples in which the best strategy depends on the environment; the bullfrogs and *Centris* bees provide examples of strategies conditional on body size or fighting ability; the treefrogs and digger wasps might be examples where different strategies are maintained in frequency dependent equilibrium.

However, in many cases it is difficult to decide which explanation is appropriate. For example in the cricket, all three hypotheses could apply. Both calling and satellite behaviour may persist because each does better in a different environment; calling may be best at times and in places where parasitic flies are scarce while satellite behaviour may be better where parasites are abundant. Alternatively, the strategies could be conditional on phenotype; maybe only the largest males or those with the most energy reserves are callers? Finally they could coexist as a frequency dependent equilibrium, each enjoying, on average, equal success. Obviously a lot of good data on the costs and benefits of calling and on individual constraints will be needed before we can say which hypothesis is correct. However even with good quantitative data it is not always easy to interpret the results. This can be illustrated by our next example.

In chapter 6 we described how a male scorpionfly, *Hylobittacus apicalis*, (also called hanging fly) must first offer a female a nuptial meal before she will allow him to copulate. Randy Thornhill (1979) discovered that males employ two methods for obtaining a prey item to give to the female. They either hunt for the prey and capture it themselves or they steal from another male. Stealing takes the form of flying straight at another male and knocking the prey forceably from his grasp or there is the rather more devious method of flying up to a male and adopting a female-like posture to try and dupe him into giving up his prize. Such 'transvestite' behaviour sometimes succeeds but sometimes the male grabs his prey item back again when he discovers that the other individual is refusing to copulate!

Individual males used both strategies to get prey; sometimes they hunted while other times they stole. Therefore success has to be measured in terms of decisions. Thornhill's data showed that stealing was more profitable than hunting; for stealing the inter-copulation time was only 17 min while for hunting it was 26 min. How do we interpret these data? There are three possibilities.

1   The pay-offs for hunting and stealing could be frequency dependent. It is obvious that all males cannot steal all the time. However, if most males were hunters then stealing would be very profitable. If there were no phenotypic constraints (e.g. body size) on whether a male could hunt or steal then we might expect frequency dependent selection to stabilise the success of the two strategies as equal. If this mixed ESS idea is correct then the data suggest that the system is not very stable. Our first possible interpretation is, therefore, that hunting and stealing represent a mixed ESS and that the data do not correctly represent the costs and benefits of the two strategies. Perhaps there is some other unmeasured cost to stealing?

2   The second possibility is that the data are accurate reflections of the pay-offs, in which case the idea that hunting and stealing are a mixed ESS must be wrong. Hunting and stealing may be part of a conditional strategy such as 'if large, steal; if small, hunt'. Maybe just as only the strongest bullfrogs defend territories, so only the best male scorpionflies go in for stealing.

3   The third possibility is that the mixed ESS idea and the data are both correct but that the population has not yet reached a stable equilibrium. Perhaps in time the frequency of stealing will increase and the pay-offs stabilise as equal. This interpretation must be considered especially as many data are now collected in habitats recently modified by man, such as polluted streams and suburban gardens. This means that some animals may be using rules adapted to different environments from the ones in which we study them.

This list of possibilities is very similar to the one given in chapter 3 (p. 66) in relation to optimality models; failure to measure some of the costs or benefits, incorrect assumptions about constraints and the possibility that animals are not well adapted.

In conclusion, although it is clear that there are several evolutionary pathways to differences between individuals within a population, there will often be problems in sorting out which hypothesis best applies to a particular example. Measurements of costs as well as benefits, together with data on individual constraints such as age and body size, are needed before we can distinguish conditional strategies from examples of a mixed ESS.

Sex change as an alternative strategy

CHANGING FROM FEMALE TO MALE

Whenever there is intense competition between males for females, it is usually the largest strongest males who will enjoy the greatest reproductive success. We have seen that one way in which young and small males can avoid direct competition with stronger rivals is to adopt sneaky mating strategies. However there is another more startling way of overcoming the disadvantage of small size when young. This is to start reproductive life as a female and then change to a male when large enough to be a successful competitor (Fig. 8.7a). This system of sex change is common in fish and is known as protogynous hermaphroditism. It will be favoured whenever an individual reproduces best as a female when small and as a male when large, and when lifetime reproductive success is greater if an individual changes sex than if it remains as either a male or female throughout its life (Ghiselin 1969; Warner 1975). As mentioned in chapter 6, the problem of sex change is closely related to that of the sex ratio. Both are part of the general problem of sex allocation.

The blue headed wrasse, *Thalassoma bifasciatum*, lives on coral reefs in the Western Atlantic. The males are brightly coloured and they defend territories on the reef. The females, which are dull in colour, choose the largest and brightest males for mating. The largest male on the reef may spawn 40 times a day at the peak of the breeding season. As we might expect from the fact that only the largest individuals are successful males, this species is a protogynous hermaphrodite; fish start reproductive life as females when small and only change to being males when large (Warner, Robertson & Leigh 1975). Sex change is socially controlled. If the largest males on a reef are removed, the next largest individuals (females) will change sex and become bright-coloured males.

The story is a little more complicated than this because there are differences in behaviour on different sized reefs (Warner & Hoffman 1980). On the largest reefs, where populations of wrasse may reach 16 000 individuals, the few largest males have potentially an enormous reproductive success. However, with large numbers of females

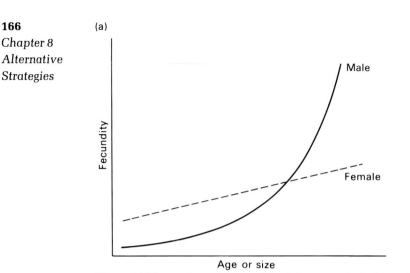

Fig. 8.7 (a) When male–male competition is intense, only the largest individuals will be successful at mating. Although female fecundity also increases with size (larger females are able to lay more eggs) the influence of male size on mating success is much stronger. Under these conditions it may pay an individual to be a female when small, because all females will breed, and a male only when large enough to be a successful competitor (after Warner 1975). An example of this case is the blueheaded wrasse (see text).

being attracted to a few males, there are also good opportunities for sneaking by small males. On large reefs some individuals are born as males and remain male throughout their lives (primary males). When small they either go in for sneaky matings, stealing into a large male's territory and attempting to spawn with the females he has attracted, or they may go round in gangs with other small males, chasing females and stimulating them to spawn with the group ('gang bangs').

On small reefs, the largest males again attract the most females but the maximum number they can attract is less than on a large reef because the total population is small, perhaps only 20 wrasse. Therefore there is less opportunity for sneaking and small males can be excluded from breeding altogether by the large males. On small reefs there appears to be an advantage for sex-changing individuals who start life as a female and then become a male when they have grown larger. Warner & Hoffman (1980) suggest that both sex-changers and primary males may be maintained in the blue headed wrasse because of different advantages on different sized reefs.

CHANGING FROM MALE TO FEMALE

More rarely, an individual may be a male when small and then change to a female when large (protandrous hermaphroditism). This type of sex change may be favoured if male–male competition is not so intense and male size has little effect on breeding success (Fig. 8.7b). An individual may then reproduce best as a male when small because

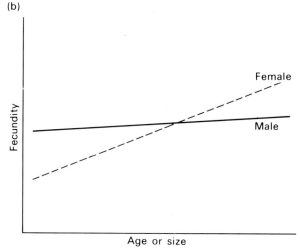

Fig. 8.7 (b) When male–male competition is less intense, female fecundity may be more dependent on body size than male fecundity. It may then pay an individual to start life as a male when it is small and change to a female when it is older and larger (after Warner 1975). An example of this case is the anemonefish, *Amphiprion* (see text).

it is able to spawn with some of the large, most fecund females.

An example of a fish that changes from male to female is the anemonefish, or clownfish, *Amphiprion akallopisos*, which lives on coral reefs in the Indian Ocean. It lives in close symbiosis with sea anemones and because there is usually only enough space for two fish to inhabit the same anemone this species lives in pairs (Fig. 8.8). In effect the habitat forces them to be monogamous. The reproductive success of a pair is limited more by the female's ability to produce eggs than by the male's ability to produce sperm and so each individual does better if the larger one is female (Fricke 1979). Like the wrasses, sex change is socially controlled. If the female is removed, the male is then joined by a smaller individual, so he changes sex and lays the eggs while the newcomer functions as a male (Fricke & Fricke 1977).

SEX CHANGE VERSUS SNEAKING

Where there is a size advantage for being one sex, the smaller individuals may sneak (as in elephant seals) or may be sequential heraphrodites, changing sex as they grow larger (as in some fish). Why do not young elephant seals start life as females and then change to males when they are large and strong, like the blue-headed wrasse? The most likely explanation is that in mammals the sexes are more differentiated than in many fish; for example there is internal fertilisation and elaborate care of the young by the female in pregnancy and during lactation. It may therefore be too costly to change sex. It is striking that all species of fish known to change sex have relatively simple sex organs and external fertilisation. Further-

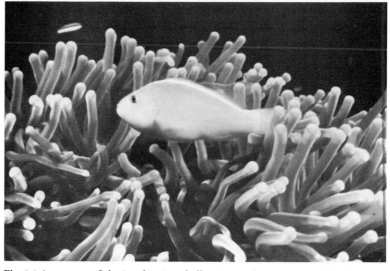

Fig. 8.8 An anemonefish, *Amphiprion akallopisos*, with its anemone. Photo by Hans Fricke.

more for a young mammal, like an elephant seal, experience may be needed to be a successful male. It may be better to forgoe reproduction when young and instead put resources into growth and learning the techniques of successful harem defence (Warner 1978).

Summary

Individuals within a species often differ in the way they compete for scarce resources such as food, mates or nest sites. Different strategies may be favoured in different environments (e.g. red and dull male sticklebacks). Sometimes behaviour is conditional on size and strength so that while the largest individuals display and fight to attract mates, smaller individuals employ sneaky strategies which make the 'best of a bad job' (eg. bullfrog satellites, hoverers in the bee *Centris pallida*). Finally alternative strategies may exist as a mixed ESS, an evolutionary equilibrium where different strategies enjoy equal success (e.g. digging and entering in digger wasps). In many cases there are not enough data on constraints and the success of the different strategies to say whether they are conditional on environment or phenotype, or an example of a mixed ESS.

Some fish change sex as a way of increasing reproductive success when small in size. Where male–male competition is intense so that only the largest individuals are successful as males, individuals may be sequential hermaphrodites, changing from female to male as they grow larger (e.g. blueheaded wrasse). More unusually, the reproductive success of a female is more size-dependent than that of a male, in which case individuals may change from male to female as they grow larger (e.g. anemone fish, *Amphiprion*).

Dawkins (1980) and Dunbar (1983) give good reviews of alternative strategies. Rubenstein (1980) discusses alternative mating strategies giving both a theoretical treatment and a review of examples. The two papers on nesting strategies in digger wasps (Brockmann & Dawkins 1979; Brockmann, Grafen & Dawkins 1979) give a clear account of the methods and problems of applying ESS ideas to alternative strategies in one species. Charnov (1979) models the optimal timing of sex change and tests his theory with data on shrimps.

Chapter 9. Co-operation and helping in birds, mammals and fish

Animals are selfish machines programmed to pass on their genes to future generations. We do not expect, therefore, to see harmonious co-operation and helping in animals; even apparently cooperative ventures such as mating usually involve selfish exploitation of one individual by another. However there is one place where we are not at all surprised to see self-sacrifice and helping, namely when parents care for their offspring by feeding, brooding and defending them against predators. This is to be expected because parents pass on their genes to future generations via offspring; by definition selection favours genes which increase in frequency from generation to genera-tion and if parental care helps to spread an individual's genes it too will be favoured by selection.

In chapter 1 we showed that it is possible to estimate the probability of a parent sharing a particular gene with its offspring. The answer for a sexually reproducing species was shown to be 0.5; in other words any gene in a parent's body has a 50 per cent chance of being duplicated in an offspring. Hence to make a genetic profit by sacrificing its life (as for example in certain arthropods such as Cecidomyian gall midges in which a mother offers herself as food for her offspring (Gould 1978)) a parent must produce more than two offspring. But in chapter 1 we also showed that full siblings have a 50 per cent probability of sharing genes, exactly the same degree of relatedness as for mother and child. Therefore if we view helping and self-sacrifice purely from the point of view of making a profit for selfish genes, our arguments for parental care should apply with equal force to caring for siblings (and also with slightly less force to caring for grandchildren and cousins). The reasons why helping offspring is commoner than helping brothers and sisters are therefore likely to be ecological and practical considerations such as the ease of recognising offspring and the degree to which offspring or siblings benefit from a given amount of aid rather than genetic factors. Some of these are summarised in Box 9.1.

Genetic predispositions and ecological constraints

Although helping in animals usually consists of parents caring for offspring, there are many examples of helping directed at other individuals. In more than 150 species of birds, for example, some individuals spend part or all of their lives helping others to reproduce: they help to feed or protect the offspring of others while apparently

172

*Chapter 9
Co-operation
and helping
in birds,
mammals and
fish*

Box 9.1 Some hypotheses about the ecological and proximate constraints that favour care of offspring over care of siblings in most species.

1 Young benefit more from a given amount of aid. One insect fed to a helpless nestling contributes more to its survival than the same insect would to the survival of a healthy sibling. In other words young are 'superbeneficiaries' (West Eberhard 1975). In species with overlap of generations this constraint may not apply since newly born younger siblings might benefit just as much from a helper's aid as would an offspring of the helper.

2 Young are generally easy to identify as relatives of the helper, while siblings may be harder to recognise as kin.

3 While young are often born close to the parents, siblings may disperse from their birth site and therefore not be available for receiving help.

4 Young, especially in slow or continuously growing species (e.g. fish), are very different in size from the helper and are therefore not likely to compete with it for food in the future. In contrast if the helper increases the survival of one of its siblings from the same brood, it may suffer in the future from increased competition.

5 Young may be more valuable in terms of expected future reproduction. If a youngster is successfully reared to maturity it has a higher expected future reproductive output than does a sibling of similar age to the helper.

foregoing the chance to breed themselves. This behaviour appears to be *altruistic*, since the helpers give benefit to others while incurring costs, such as expenditure of energy and risk of predation, to themselves (chapter 1). In this chapter we will discuss how this kind of helping might be favoured by selection in birds, mammals, and fish. The problem can be considered as having two parts.

1 *The genetic predispositions for helping*: perhaps helpers substitute sibling care for parental care, in which case helping could be viewed as a form of aid to close relatives or *nepotism*. As discussed in the last paragraph it is possible that as great, or even greater genetic pay-offs can be gained by sibling care as by parental care.

2 *The ecological constraints which favour helping in some species but not in others*: as Box 9.1 shows, parental care is normally favoured over sibling care because of practical and ecological constraints. Perhaps some of these constraints are reversed or absent in species with helpers.

We will start by describing one of the best studied examples of helping in birds and then go on to consider some of the complications which arise in trying to extend the results of this study to other species.

An example of helping in birds—the Florida Scrub Jay

173

Chapter 9
Co-operation
and helping
in birds,
mammals and
fish

The Florida scrub jay (*Aphelocoma coerulescens*), as its name implies, lives in oak scrub in Florida. Suitable habitat for birds is patchy and scarce, so there are isolated pockets of breeding jays with totally unoccupied and unsuitable areas between. The birds live all the year round in territories, each of which contains a group of one breeding pair and between 0 and 6 helpers; these birds aid the parents in rearing the nestlings (Fig. 9.1). About three quarters of all helpers are offspring of the breeding pair from a previous season: if both parents have survived they are therefore full siblings of the chicks they help to raise, if one parent has died they are half siblings. Most offspring spend one or more years helping on their parents' territory and by the age of 3–5 years they have usually left home to start breeding themselves. The scrub jay has been studied for over ten years by Glen Woolfenden (1975) and his beautifully detailed records reveal a number of clear patterns which show how genetic and ecological factors favour helping.

1 *Breeders benefit from the presence of helpers.* Breeding groups of scrub jays with helpers rear more young than those without help (Table 9.1). Helpers make two kinds of contribution

(a) They help to defend the nest against predators such as snakes by mobbing and giving alarm calls to warn the chicks. This antipredator behaviour is primarily responsible for the increase in survival of chicks when helpers are present in a group.

(b) They help to feed the nestlings, providing up to 30 per cent of all food brought to the chicks in some groups. While this apparently does not result in an increase in chick survival since the total amount of food brought to the nest is not increased by the efforts of helpers, it does relieve the burden on parents so that they survive better from one

Fig. 9.1 The Florida scrub jay. Helpers provide food for the young in the nest and defend the nest against predators such as snakes (photo by Glen Woolfenden).

174

Chapter 9

Co-operation

and helping

in birds,

mammals and

fish

season to the next. Breeders with helpers have to put less effort into feeding their chicks and probably as a consequence of this their annual survival rate is 87 per cent, as opposed to 80 per cent for breeders without helpers. Since the helpers and parents are related to one another by about 0.5, this increase in parental survival represents a genetic gain for helpers in addition to that made by rearing younger siblings.

Table 9.1 Breeding pairs of Florida scrub jays benefit from the presence of helpers. Both experienced and inexperienced (first time) breeders rear more young when they have helpers (data of Woolfenden analysed by Emlen 1978).

	No helpers	With helpers	Average number of helpers
Inexperienced pairs	1.03	2.06	1.7
Experienced pairs	1.62	2.20	1.9

2 *Habitat saturation is an ecological constraint.* Do helpers benefit from staying at home and helping? Clearly they make some genetic profit because they help to rear their own younger siblings which, as we said at the beginning of the chapter, is approximately equivalent to rearing offspring. But the crucial question is whether helpers could in theory do better by setting up their own territory and breeding themselves right from the start, or by staying at home and helping. A helper adds on average 0.32 offspring to its parent's production of young, and a novice pair of breeders rears about 1.36 young (a generous estimate of what a helper would do if it set off and bred instead of helping). So a helper could make 0.68 genetic equivalents by going off to breed instead of only 0.16 equivalents by staying at home, assuming both siblings and offspring are related to it by 0.5 (Table 9.2). These calculations are greatly oversimplified. They do not, for example, take into account the following: the mortality of young birds may be increased by setting off too early in life to seek a territory; by staying at home a youngster can increase the survival of its parents,

Table 9.2 A scrub jay helper would do better by setting up its own breeding territory if it could find one, than by helping its parents at home (Emlen 1978).

Option		Result
1 Stay at home and help	Young produced by experienced parents with no help	1.62
	Young produced by pair with one helper	1.94
	Extra young due to presence of helper	0.32
	Genetic equivalents to helper (r = 0.5)	0.16
2 Go off and rear own young:	Young reared by first time breeders	1.36
	Genetic equivalents (r = 0.5)	0.68

175

Chapter 9
Co-operation
and helping
in birds,
mammals and
fish

which as mentioned earlier is part of the genetic pay-off for helping; helping may act as a form of training and increase the youngster's success when it breeds on its own; and the best strategy for a young bird may depend on what all the others do (the more they decide to leave home the more competition there will be for breeding sites). But these complications are not likely to alter the general conclusion that a young scrub jay would do better to leave home if it could find a territory, in spite of the fact that it rears siblings when it helps at home. The most plausible interpretation is that helpers stay at home because there are no spaces for them to set up a breeding territory: the habitat is saturated. This interpretation is supported by Woolfenden's observation that as soon as vacancies arise helpers usually leave home to set up on their own.

3 *Males benefit more than females from helping.* One of the commonest ways for a young male scrub jay to acquire a breeding space is by inheriting part of its parents' territory (Woolfenden & Fitzpatrick 1978; Fig. 9.2). The mortality rate of adult males is low and

Fig. 9.2 An illustration of how male scrub jays may inherit part of their parents' territory. This is one of the benefits of helping. The diagrams show how three helpers (B,C,D) acquired territories. Shaded areas represent land inherited by two sons (B,C) and one grandson (D) of the pair occupying territory A. Broken lines show the incipient point of budding off and small arrows show how territorial boundaries expanded (from Woolfenden & Fitzpatrick 1978).

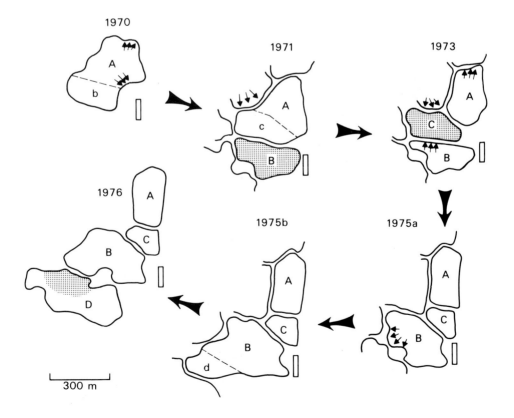

176

Chapter 9
Co-operation
and helping
in birds,
mammals and
fish

therefore very few vacant territories appear each year; a young male's best chance of acquiring a territory is to help his parents to increase the size of their territory until it is large enough for him to split off part of it. By helping his parent the young male increases their production of young and therefore the size of group occupying the territory. Larger groups can expand their territories at the expense of small groups, and eventually the young male buds off to set up his own territory at the edge of the parental area. Therefore young males benefit in two ways from helping their parents. They propagate their genes by rearing siblings and they create an 'army' which will help to expand the parents' territory until it is large enough to provide space for them to set up on their own. When there are several male helpers in a territory there is a dominance hierarchy and the oldest dominant male is the first to inherit his own territory.

As with most birds, females do not breed in or near their parents' territory, but instead they disperse to find a vacant breeding space elsewhere. As discussed in chapter 7 this sex difference in dispersal might be a mechanism to avoid inbreeding, since females usually end up mating with an unrelated male as a consequence of dispersal. The

---

Box 9.2 Genetic predispositions for males to become helpers.

Ric Charnov (1981) has pointed out that there may be a genetic predisposition for males to become helpers. This arises from uncertainty of paternity. Suppose that a male does not father all his mate's offspring (because he is cuckolded). A daughter of this male is more closely related to her own children than to her siblings. Her relatedness to her children is always 0.5, but if she and her siblings are not all fathered by the same male the average relatedness between her and her siblings is less than 0.5. In the extreme example where each offspring has a different father, the degree of relatedness between brothers and sisters is 0.25.

Now consider the effect of the same situation on a male. If he never fathers his mate's children his relatedness to them is zero, while he is always related to his siblings by at least 0.25 (since all siblings share the same mother). If males father half their mate's children their average relatedness to offspring is $0.5/2 = 0.25$. Their relatedness to siblings in this case is 0.25 genes shared via the mother) + 0.25/2 (genes shared by siblings through the father). This gives a total of 0.375: again males are more closely related to their siblings than to their own mate's offspring.

These calculations serve to illustrate the point that any degree of uncertainty of paternity should predispose males to become helpers and rear siblings instead of offspring.

female's benefit from helping is therefore less than that of the male: she rears close relatives and has a safe base from which to explore, but she does not inherit her own breeding territory as a result of helping. As one might predict from these observations, females help less than males, and among males the older more dominant birds help most. In other words help is given to the parents in proportion to the future expected gain for the helper. In addition to the ecological factors predisposing males to help more than females, there may possibly be genetic considerations (Box 9.2).

To summarise, the three main conclusions of the scrub jay are (a) helpers are usually very close relatives of the nestlings and are therefore genetically predisposed to provide aid (b) the crucial ecological constraint is shortage of breeding habitat which prevents young birds from setting up their own territories and (c) males help more than females because they benefit more in the future, by gaining a territory (Fig. 9.3)

177
Chapter 9
Co-operation
and helping
in birds,
mammals and
fish

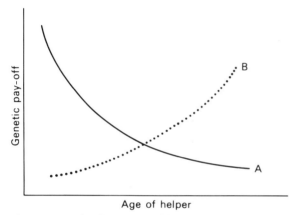

Fig. 9.3 A graphical summary of helping in the scrub jay and many other similar species. The genetic pay-off for helping at home (line A) decreases with increasing age because it is likely that eventually one or both parents will be replaced and the degree of relatedness between helpers and young is therefore decreased. The benefit from breeding (line B) increases with age because as birds get older they are more likely to take over their own territory. The position of the curves is influenced by the ecology of species. For example when territories are plentiful, curve B is shifted far to the left and helping may not occur at all, but when territories are scarce the period of helping is prolonged because curve B moves to the right. The two curves are not necessarily independent, since helping might increase the experience of young birds and therefore their future breeding success.

Helping in other species and some complications

The pattern found by Woolfenden in scrub jays seems to apply to many bird, mammal, and perhaps fish species with helpers. Helpers are usually offspring of the breeding pair, their presence usually increases the breeding success of the group, and they usually have little opportunity to breed themselves because of ecological constraints. The black-backed jackal (*Canis mesomelas*) in the Serengeti

178

Chapter 9
Co-operation
and helping
in birds,
mammals and
fish

fits this pattern. Monogamous breeding pairs have 1–3 young from previous litters that act as helpers by regurgitating food for the pups and lactating female as well as grooming, guarding and playing with the pups (Fig. 9.4, Fig. 9.5). Similarly in the Princess of Burundi cichlid fish (*Lamprologus brichardi*) helpers are young of the breeding pair from previous broods and they guard the eggs and larvae against predators (Taborsky & Limburger 1981).

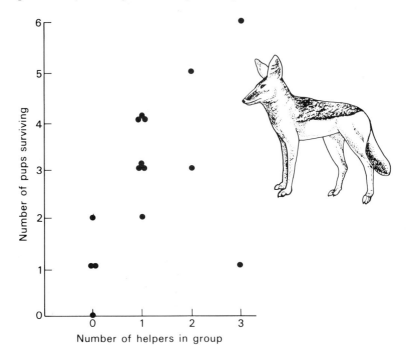

Fig. 9.4 Breeding success of black-backed jackals is positively correlated with the number of helpers in a group (Moehlman 1979).

DO HELPERS REALLY HELP?

A correlation between the presence of helpers and breeding success of the group does not necessarily show that helpers help. The correlation could arise, for example, because good quality parents produce many young each year and therefore have both a high annual success and a large number of helpers from previous seasons. Alternatively some pairs might live in good territories and have enough food or shelter from predators to rear many young each year and therefore acquire a large number of helpers. In order to disentangle cause and effect it is necessary to do an experiment such as the one carried out by Jerram Brown *et al.* (1978, 1981) on grey-crowned babblers (*Pomatostomus temporalis*). The grey-crowned babbler lives in year round territorial groups of 2–13 birds in open woodland in Queensland, Australia. As with the scrub jays, each group consists of a breeding pair and a variable number of helpers which are young from previous broods. The

179

Chapter 9
Co-operation
and helping
in birds,
mammals and
fish

breeding success of a group is positively correlated with the number of helpers present, and in order to show that helpers actually cause the increase in success Brown did a removal experiment. He removed all the helpers but one from 10 breeding groups and left another 10 groups as controls with 4–6 helpers per group. The control groups reared more young per season than the experimentals, mainly because the helpers feed the young and relieve the pressure on the breeding female so that she can recover from one breeding attempt and start again sooner.

HELPERS ARE NOT ALWAYS RELATIVES

The scrub jay pattern does not, however, apply to all species with helpers. In particular helpers may be quite unrelated to the offspring they help to rear. The dwarf mongoose (*Helogale parvula*) is an example. It is a small diurnal carnivore that lives in groups of about ten individuals made up of a breeding pair and helpers; the helpers are either offspring of the breeding pair or they are unrelated immigrants. Females tend to be more active as helpers and they bring food such as beetles, termites and millipedes to the young in their den as well as guarding the den against predators such as other species of mongoose (Rood 1978). The striking observation made by Rood is that unrelated helpers do as much helping as those which are siblings of the young ones in the den. In one troop an immigrant female did more work than any other individual, including the father and mother! In contrast to related helpers, unrelated individuals gain no genetic profit through the young ones they help to raise, so the only possible gain they can make is to increase their own success in the future. In the mongoose this comes about because unrelated helpers, especially females, sometimes take over as breeders when one of the resident parents dies. As with the scrub jays there is a shortage of breeding opportunities and helping is part of the strategy of eventually acquiring a slot as a breeder. But why should the unrelated mongoose help instead of just waiting on the sidelines for a chance to start breeding? There are three factors which might be important:

1 Helping may be a form of 'payment' for permission to stay on a breeding territory while waiting to take over. If the resident breeding pair gain nothing from having an unrelated helper close at hand they may simply evict it.

2 Helping keeps the group and therefore the territory intact, which is vital to the helper's future success when it becomes a breeder itself.

3 Finally, some of the young reared by the helper will in turn help it when it becomes a breeder, regardless of whether or not they are relatives. These points can be summarised by saying that helping is a long term investment for selfish future breeding success.

A similar interpretation may be applicable to helpers in the anemone fish *Amphiprion akallopisos*. Breeding pairs of fish defend

(a)

(b)

Fig. 9.5 Black-backed jackals (*Canis mesomelas*) are monogamous and have helpers who are offspring from a previous litter. (a) A mated pair grooming their year-old son (centre) who remains on the parental territory as a helper. Helpers contribute to pup survival by regurgitating food (b) chasing off predators (c) in this case a spotted hyena, and by helping to defend the territory (d) the intruder is on the right (Photos by © Patricia D. Moehlman 1980).

(c)

(d)

anemones, which are an essential scarce resource for successful reproduction. Sometimes the pair are helped in defence by a non-breeder, which is most unlikely to be a relative of the breeding pair since the young fish disperse in the plankton before settling on a territory. Instead the helpers are probably unrelated and as hypothesised for mongoose helpers, they may be 'hopeful reproductives' waiting for a chance to take over when one of the breeders dies (Fricke 1979). Helping may again be a form of payment for permission to stay in the territory.

Chapter 9
Co-operation
and helping
in birds,
mammals and
fish

The idea of helping as a payment for permission to remain in a group is clearly illustrated by the pied kingfisher (*Ceryle rudis*) (Fig. 9.6) in which the breeding pair apparently accept unrelated helpers only when the payment is sufficiently advantageous to them. The pied kingfisher, unlike most of the species referred to so far, is a colonial nester and does not live in a year round territory. Uli Reyer (1980) compared two colonies, one at Lake Victoria and the other at Lake Naivasha in East Africa. At Naivasha breeding pairs have at most one helper which is always a male offspring of the breeding pair (a 'primary helper'). He helps by feeding fish to his mother, defending the nest against predators, and bringing food to the nestlings. At Lake Victoria there are sometimes secondary as well as primary helpers; these birds are also males but are unrelated to the breeding pair. They show up at both colonies just after the young hatch out and try to feed the breeding female of a pair; they are persistently chased away by the breeding male at Naivasha, but at Victoria they are eventually tolerated and allowed to stay and help to feed the young. The reason why secondary helpers are accepted at one lake but not the other is

Fig. 9.6 A pied kingfisher bringing food to the nest. (Photograph by Uli Reyer)

183
Chapter 9
*Co-operation
and helping
in birds,
mammals and
fish*

probably related to feeding conditions. Victoria is not as good for fishing as Naivasha: the water is rougher so that it takes longer for a kingfisher to catch a fish, the fish are smaller, and the feeding grounds are further away from the colony than at Naivasha. This means that parents with only one helper have difficulty in bringing enough food for their chicks and the secondary helpers make a big difference to breeding success (Table 9.3). At Lake Naivasha one helper is enough to rear all the young successfully and a secondary helper could therefore contribute little to the success of a breeding pair.

Table 9.3 Helpers in the pied kingfisher. At Lake Naivasha nests have no helpers or one helper, feeding conditions are good and one helper is therefore enough to ensure that all the chicks get enough to eat. At Lake Victoria there are sometimes second (unrelated) helpers, feeding conditions are poor, and the second helper makes a big difference to chick survival (Reyer 1980).

	Per cent nests With helpers 0 1 2	Feeding conditions Time to catch a fish (min)	Per cent dives successful	Clutch size	Breeding Success Young hatched	Young fledged with helpers 0 1 2
Lake Naivasha	72 28 0	5.9	79	5.0	4.5	3.7 4.3 —
Lake Victoria	30 50 20	13.0	24	4.9	4.6	1.8 3.6 4.6

The conclusion is that unrelated helpers are accepted only when their help is effective at increasing the breeder's success. But why should kingfishers help at all? There is no shortage of breeding sites since the birds make a hole in a bank for nesting and there is plenty of bank space, but instead mates are the scarce resource which prevents males from breeding. Females are in short supply, probably because they suffer heavy mortality on the nest due to predation by lizards, snakes and mongooses (the sex ratio among adults is about 1.8:1). All the kingfisher helpers are males without a mate. Primary males stay at home and make some genetic profit by rearing younger siblings until they can set off with a mate of their own. Secondary helpers may also be on the look out for a female, but they are waiting to take over the female of the pair they are helping. More than half the secondary helpers return to become breeders in the following year, and of these half pair with the female they had previously helped. Presumably take overs normally occur when the resident male dies, but there is also a chance that the helper may cuckold or displace the breeding male. This is why the male only accepts a helper if there is an immediate large benefit, which has to be weighed against a possible future cost.

We can now extend the conclusions of the scrub jay study by saying that helping in birds, mammals and fish may lead to genetic gain through aid to relatives or through future reproduction of the helper (relatedness is not essential for helping to be favoured by selection) and the environmental constraint which leads to helping might be a shortage of territories or mates.

184

*Chapter 9
Co-operation
and helping
in birds,
mammals and
fish*

Conflict in breeding groups

We have already seen in the pied kingfisher that helpers and breeders may have conflicting interests. It has even been suggested that helpers in some species may try to disrupt breeding in order to decrease the number of future competitors and therefore increase their own chances of success at getting a territory (Zahavi 1974), although there is as yet no firm evidence for this view. However conflict of interest is very clearly demonstrated in birds where co-operative breeding involves several females sharing the same nest (as opposed to the species described so far in which just one female breeds).

OSTRICHES

In the ostrich (*Struthio camelus*) as many as six females share a nest. The male defends a large territory (14 km²) and builds a nest which consists of a scrape in the ground. He then acquires a mate (called the 'major female') and she starts to lay eggs. After a few days other females which may or may not have mated with the resident male ('minor females') start to add their eggs to the nest. The minor females take no part in incubating and guarding the eggs, this is done entirely by the major hen and her mate. The minor females are therefore in effect brood parasites. The nest usually ends up containing about 30–40 eggs, which is too many for one bird to incubate, so some eggs are pushed to the edge of the nest where they fail to hatch because they get overheated in the sun or destroyed by predators (Fig. 9.7). Many

Fig. 9.7 Communal breeding in ostriches. A major hen stands guard over her nest. She has rolled the eggs of minor hens to the edge of the nest where they will succumb to predators or overheat in the sun.

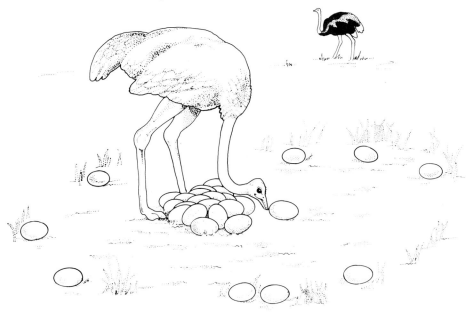

observers have noted that the incubating female pushes eggs to the edge of the nest but until recently its full significance had not been appreciated. Sauer & Sauer (1959) viewed it as a method of regulating the temperature of eggs, and the local natives in East Africa thought it was a method of protecting the nest against fire. Their idea was that when a fire swept across the savannah the eggs at the edge of the nest would explode in the heat and dampen the ground enough to prevent the fire from destroying the nest!

185

*Chapter 9
Co-operation
and helping
in birds,
mammals and
fish*

The real explanation of egg rolling, discovered by Brian Bertram (1979, in prep.) is no less bizarre. The major hen can recognise her own eggs, perhaps by weight or by surface pore patterns (the eggs are plain white but have a very bumpy surface), and she pushes the eggs of minor females to the edge of the nest while keeping her own in the middle. In a typical nest 100 percent of the major female's and about 50 percent of the other eggs are incubated. The major female does not try, however, to prevent minor females from laying their eggs in her nest. In fact she politely gets off the nest and stands aside to allow them to lay. The hens sharing a nest are not close relatives, so the major female is not making a genetic gain by rearing nieces, nephews and cousins. Instead she probably gains because she protects her own eggs by surrounding them with others. An ostrich's nest is an attractive meal for predators, equivalent in size to about 900 hens' eggs! However large clutches are not proportionately more vulnerable to predators, and predators such as jackals and vultures do not normally eat more than a few eggs during an attack. The major female's eggs therefore benefit from being mixed in with and diluted by those of other hens (chapter 4). The chance that one of the major female's eggs will succumb to a predator is further reduced by the presence of other eggs because her eggs tend to be in the middle of the nest and predators probably take the first egg they encounter at the edge of the clutch. It appears therefore that the major female actually benefits by having minor hens' eggs in her nest, but this does not explain why she incubates them. She can discriminate between her own and others' eggs and could therefore presumably make sure that only her own hatch. One hypothesis is that the same dilution argument can be applied to newly hatched chicks and this is why the major hen incubates more than her own brood (chapter 4).

What about the minor females? They are often females without a mate or ones that have lost their own clutch (a major hen in one territory may therefore become a minor one elsewhere). They therefore have no chance to be major hens and they try to get a few offspring into the next generation by parasitising major hens. The communal behaviour of ostriches can therefore be viewed as intraspecific nest parasitism in which the 'hosts' happen to be able to turn things to their own advantage.

In addition to the obvious conflict of interests between females, there may also be conflict between male and female. The male fathers

186

Chapter 9
Co-operation
and helping
in birds,
mammals and
fish

some of the minor female's young so he is equally interested in the survival of their chicks, while the major female is only interested in the survival of her own. It is not known what effect this conflict has on the reproductive behaviour.

GROOVE-BILLED ANIS

The groove-billed ani (Crotophaga sulcirostris) is another species in which several females share the same nest. The ani lives in a group territory defended by 1–4 monogamous breeding pairs. The group builds a communal nest in which all the females lay some eggs and to which all the group members contribute parental care. As with the shared nests of ostriches, there are too many eggs to incubate successfully and females compete to make sure that their eggs are the ones which survive. Competition takes the form of females rolling each other's eggs out of the nest, so that by the time incubation starts there are usually broken eggs strewn over the ground below the nest. Female anis are apparently unable to recognise their own eggs so each female rolls eggs out of the nest only up to the time when she herself starts to lay, otherwise she would run the risk of evicting her own eggs. Sandy Vehrencamp (1977) found that there is a dominance hierarchy among the females in a group and that the most dominant female ends up with the largest number of eggs in the nest at the time of incubation. She achieves this by starting to lay after the others, who are therefore unable to evict her eggs. The subordinate females start to lay earlier than the dominant, but they have several tactics for increasing the chances of survival of their eggs: they lay more eggs than dominants, usually wait two or three days between eggs, often produce a late egg after incubation has started, and initiate incubation earlier than the dominant. Once incubation starts the embryos begin to develop and the dominant female is forced to stop laying since otherwise her young would be late hatchers and at a disadvantage in competition between nestlings for food. In spite of these tactics, nearly all the eggs of dominant females survive to be incubated while less than half those of subordinates survive (Fig. 9.8). Although the existence of conflict between females is clear it is not yet known how dominant females succeed in laying last.

There is also an asymmetry in the extent to which female anis care for the eggs and young. Surprisingly the dominant female, with the biggest stake in the nest, does the least care (although the dominant male is very attentive as a parent). Again it is not known how the asymmetry is brought about, although presumably the subordinate females would not care for young if the nest contained none of their own offspring. (In another communal nester, the acorn woodpecker Melanerpes formicovorous, individuals that have not contributed to the nest show no parental care (Stacey 1979).)

187
Chapter 9
Co-operation
and helping
in birds,
mammals and
fish

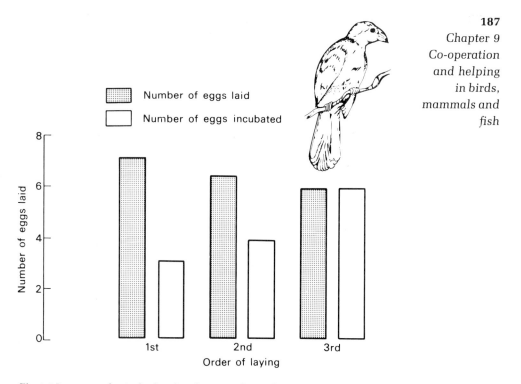

Fig. 9.8 In groups of anis the first female to start laying lays more eggs but has fewer that survive to be incubated than the last female (Vehrencamp 1977).

The ecological condition which favours communal nesting in anis seems to be, as with ostriches, predation pressure. In open pasture habitats where the nests are conspicuous to diurnal predators, groups are better able to defend the nest against predators at least in the egg stage. In dense marsh habitats the incubating anis suffer a great deal of predation at night and when the incubation is shared between the individuals of a group each one has a higher chance of survival (Vehrencamp 1978). However both these advantages are benefits to the average group member and as we have seen, dominant birds gain much more out of being in a group than do subordinates. Subordinates must still gain more from group nesting in terms of predator defence than they lose from interactions with dominants, or they would leave the group and nest as solitary pairs.

Conclusion

The major points to emerge from studies of helping in birds, mammals and fish are as follows.
1 Production of offspring is only one pathway to genetic representation in future generations; helping to rear siblings is the commonest alternative route in vertebrates.
2 Ecological conditions such as shortage of nesting territories or

188

Chapter 9
Co-operation
and helping
in birds,
mammals and
fish

mates may limit the options open to a young individual so that it has to help rear siblings instead of its own young.

3 It may even pay to help unrelated breeders as a form of investment for future breeding opportunities. Whether or not breeders accept the help may depend on the benefits they receive.

4 When ecological conditions such as severe predation pressure favour shared nests by a group of females, the interaction within the group can best be understood in terms of conflict between individuals rather than as adaptations for communal breeding.

dominant vs. subordinate

Summary

Selfish gene theory predicts that individuals may help close relatives as a means of passing on genes to future generations. The commonest form of helping is parental care: although caring for siblings is usually genetically equivalent to caring for offspring, ecological and practical considerations favour parental over sibling care. In some birds, mammals, and fish help is given to younger siblings by individuals that have not yet started to breed themselves. Helping is often imposed by ecological constraints such as a shortage of territories or mates which prevent helpers from breeding themselves. Helpers gain not only by aiding close relatives but also by increasing their own breeding prospects for the future. Sometimes helpers are unrelated to those they help, and they obtain only the second of these two benefits. In some birds several females share the same nest but the benefits of communal nesting, probably defence against predators, are not shared equally between the group. Dominant females do much better than subordinates.

Further reading

Emlen (1978) and Brown (1978) review work on helpers at the nest in birds.

Vehrencamp (1980) is a good review of helping in birds, arthropods and mammals in which the similarities and differences between insects and vertebrates are discussed. Gaston (1978) develops the idea that helpers 'pay' for permission to stay on the territory.

Watts & Stokes (1972) describe how brothers co-operate in competition for mates in the wild turkey.

Chapter 10. Co-operation and altruism in the social insects

The social insects

THE PROBLEM

Co-operation and helping in vertebrates pales into insignificance beside the social insects. In these insects apparent self-sacrifice reaches the point where some individuals are completely sterile; they never reproduce themselves but instead spend their whole adult lives devoted to rearing the young of others. As Darwin himself and many other biologists since his time realised, this presents a real paradox, for if natural selection favours maximum genetic contribution to future generations, how can it lead to the development of totally sterile individuals that never reproduce? In the last chapter we saw that self-sacrifice and helping others to breed may be favoured by selection in vertebrates because the help is usually given to close relatives. Some individuals are prevented from having their own offspring by ecological constraints such as lack of a territory, so instead they help to rear their younger siblings. We also saw that helping non-relatives or relatives is sometimes part of a long-term strategy by which an individual eventually gains a mate or a territory. In this chapter we will consider to what extent similar ideas can be used to understand how sterile castes and helping have evolved in the social insects.

THE DEFINITION OF 'SOCIAL INSECT'

What exactly is meant by the term 'social insect'? To be more precise, this chapter is largely about the *eusocial insects*, which are characterised by three features: they have co-operative care of the young involving more individuals than just the mother; they have sterile castes; they have overlap of generations so that mother, offspring and grandchildren are all alive at the same time. This is important because it provides the opportunity for the young to rear their younger siblings instead of having their own offspring, as in the scrub jays described in the last chapter. All three of these traits are necessary for a species to qualify as eusocial, but there are many species with intermediate stages, such as co-operative nest building without sterile castes. Eusociality occurs in three insect orders: Hymenoptera (ants, bees and wasps), Isoptera (termites) and Homoptera (aphids). The existence of eusociality in the first two orders has been well known for a long time, but in aphids it has been discovered only fairly recently (Aoki 1977).

190

Chapter 10
Co-operation
and altruism
in the social
insects

As we shall see later on, there are special genetic predispositions towards the evolution of sterile castes in Hymenoptera and aphids, but not in termites.

The social insects are not only important because of their central role in the attempts of evolutionary theorists to understand the origin of altruism, but they are also extremely impressive in terms of their natural history. In a persuasive piece of salesmanship E. O. Wilson (1975) advertises that there are more than 12 000 species of social insects in the world, which is approximately equivalent to all the known species of birds and mammals. The staggering natural history of social insects can be illustrated by the following small sample of facts. In terms of size, a colony of African driver ants (*Dorylus wilverthi*) may contain up to 22 million individuals weighing a total of 20 kg. In terms of communication, the honey bee dance language, in

Fig. 10.1 Examples of castes in social insects. The top row are the female castes and male of the myrmicine ant, *Pheidole kingi instabilis*. (a) minor worker, (b) media worker, (c) major worker, (d) male and (e) queen. The bottom row are various specialised castes in other species. (f) Soldier of the ant *Camponotus truncatus* blocking a nest entrance with its plug-like head which serves as a 'living entrance' to the nest. (g) A sterile caste of the nasute termite, *Nasutitermes exitiosus* which has a head shaped like a water pistol to spray noxious substances at an approaching enemy. (h) Replete worker of a *Myrmecocystus* ant, which lives permanently in the nest as a 'living storage cask' (from Wilson 1971).

(a)

(b)

(c)

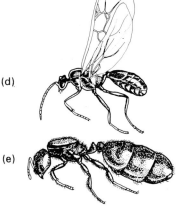
(d)

(e)

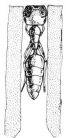
(f)

(g)

(h)

191

*Chapter 10
Co-operation
and altruism
in the social
insects*

which successful foragers tell other worker bees about the direction and distance to a food source, is one of the only known communication systems in wild animals in which an abstract code (the speed and orientation of the dance) is used to transmit information about remote objects. In terms of feeding ecology, the diets of social insects include seeds, animal prey, fungus grown in special gardens on collections of leaves or caterpillar faeces, and the excreta ('honeydew') of tended herds of aphids. Social insect colonies are often populated by individuals specialised to perform different tasks (so called castes). Sometimes castes have bizarre morphological modifications to help them carry out their jobs (Fig. 10.1). For example the head of soldier termites of the species *Nasutitermes exitiosus* is modified into a 'water pistol' used for squirting defensive sticky droplets at enemies, while the head of soldiers in ant species such as *Camponotus truncatus* is shaped like a plug which fits neatly into the nest entrance to keep out intruders, rather like the operculum of a snail.

In the following sections, we will first describe the life history of one example of a eusocial insect to provide a background, then consider two theories which have been put forward to account for the evolutionary origins of sterile castes before going on to discuss the special features of hymenopteran genetics, which are thought to predispose this group to evolve sterile castes.

The life cycle and natural history of a social insect

Myrmica rubra is a species of ant commonly found in Europe in woodlands, farmland and gardens. It builds a nest, the chambers of which are excavated in the ground underneath flat stones or sometimes in rotten tree stumps or in open soil. The nest is started by a single fertile queen. She is fertilised during a 'nuptial flight' in August or September in which very large numbers of winged reproductive females and males swarm around in the air and copulate (only the sexual forms can fly, and then only at this stage in their life cycle). The queen loses her wings after the nuptial flight and spends her first winter sealed inside the nest chamber which she has built by excavating a hole in the ground or in a tree stump. During the following summer, eggs which she has laid develop into larvae and may mature into adult workers (the typical ants one sees scurrying around near ants' nests) before the autumn or in the following spring. Up to the time of maturity of the first workers, the queen feeds herself and her brood entirely off her own reserves of fat and protein, but when the workers mature they start to care for their younger siblings and collect food for them. The workers are female, but they are sterile. They never develop wings, their ovaries do not mature and they never take part in a nuptial flight. In successive years the colony and its nest grow slowly, until after about nine years it contains roughly 1000 workers and still a single queen who has laid all the eggs. At this stage

192

Chapter 10
Co-operation
and altruism
in the social
insects

colony starts to produce a new generation of reproductives, winged females and males that eventually leave the colony on their nuptial flights. The old colony may then continue for a few more years, but as soon as the old queen dies and stops laying eggs to replenish the worker force, the colony dwindles in size and dies off.

This picture of the life cycle is typical of many temperate zone ants, although of course the details vary greatly from species to species. The details of worker behaviour are also very variable, but the following generalisations are typical of many species. Workers usually spend the first few weeks of their lives inside the colony handling dead prey which have been brought back by foragers, feeding the larvae and the queen with regurgitated food, cleaning the nest, and guarding the entrance. Later in life (this change takes place at an age of about 40 days in *Formica polyctena*, a species which has been studied in detail) workers begin to do jobs outside the colony, mainly foraging and defence against enemies. The total length of life of ant workers is not very well known but probably ranges from a few weeks to a few years. In wasps and bees, workers usually live for about 3–10 weeks. In addition to changes in worker behaviour with age, in many species of ants there are two major castes of worker (both sterile females): soldiers and normal workers. Soldiers are usually larger and have large heads with jaws or glands for producing defensive secretions. As their name implies they are specialists in colony defence.

The females belonging to different castes (queen, worker, soldier) do not usually differ genetically: the determination of caste depends on environmental conditions during larval development. In *Myrmica*, for example, whether a larva develops into a queen or a worker depends on factors such as nutrition, temperature and age of the queen who laid the egg. In honeybees the queen can suppress the development of new queens by chemical signals which prevent the workers from feeding larvae the special diet ('Royal Jelly') needed to make them grow into queens.

How eusociality evolved: two theories

In this section we will reconstruct two possible pathways along which the evolution of sterile castes proceded. The theories described refer to evolutionary history and therefore cannot be tested by direct experiments. Instead they have to be assessed in terms of the feasibility of the selective pressures they propose and by how well they account for the variety of stages of development of eusociality seen in present day insects.

In trying to understand how eusociality and in particular sterile castes evolved, it is useful to remember the distinction made in the last chapter between *ecological constraints* and *genetic predispositions*. The ecological constraints are the features of the environment which may either favour group living and co-operative breeding and/or

193

Chapter 10
Co-operation
and altruism
in the social
insects

reduce the chances for young individuals to breed themselves. The genetic predispositions refer to the degree of gene sharing between helpers and those they help: the greater the degree of relatedness, the smaller the benefit: cost ratio that is necessary for helping to be favoured by selection (chapter 1). In the social insects members of a colony are very often part of the same family, so that relatedness is high and workers are genetically predisposed to help. There are two theories about how sterile castes have evolved in insects, both of which postulate certain ecological constraints and genetic predispositions (Fig. 10.2).

(a) **Families at home**

Solitary parasitoids

Nest guarding by ♀

Young stay at home
and help defend nest
or enlarge it

Young permanently at
home and never breed

(b) **Shared nests**

Several reproductive females
(often sisters) build nests
close together

Co-operation in nest building
and defence, but each female
reproduces herself

Domination by one ♀ in group.
Others lose chance to reproduce

Overlap of generations and
young females becoming workers

Fig. 10.2 Two hypotheses about the origin of eusociality in insects. According to hypothesis (a), sterile castes originated as daughters staying at home to help their mothers; according to (b), they originated from groups of reproductives nesting together in which one female dominated the rest.

STAYING AT HOME TO HELP

(a) Ecological constraints

The ancestors of present day social insects were probably parasitoid wasps which laid their eggs inside or on the surface of a host, on which the larvae fed as they grew up (Evans 1977). A simple form of parental care in present-day parasitoids is to anaesthetise a host by stinging it, and then to carry it back to a safe site such as a hole in the ground or crevice in a tree before laying an egg on it. A slightly more advanced form of parental behaviour, seen for example in the digger wasp (chapter 8) is to create a safe place by burrowing into the ground. The burrow is then provisioned with one or more paralysed insect prey and an egg laid on each before sealing up the burrow entrance.

194

Chapter 10
Co-operation
and altruism
in the social
insects

Parental care of this type could lead to the formation of larger groups if there was sufficient ecological pressure for the offspring to stay near their birth site after reaching maturity and for mothers to tend their broods. Two main ecological factors are likely to have favoured staying at home.

1 Defence of the eggs and larvae against parasites. A major cause of mortality of young in provisioning species such as digger wasps is often parasitism by other insects, and in eusocial species one of the important tasks of workers is to defend the nest against parasites and other enemies.

2 Nest building. Although the ancestral parasitoids probably carried prey to naturally occurring refuges such as cracks and holes (or even burrows dug by their hosts), this soon gave rise to the construction of artificial refuges known as nests. In habitats where natural sites were in short supply, there would have been strong selection for the ability to build nests out of mud, chewed vegetation, and so on. Nest building is a laborious and time-consuming business, so it is easy to imagine that newly matured adults might have done better to stay at home and help their mother to enlarge or repair her nest rather than go off to start on their own. At first they may have simply used their mother's nest for laying their own eggs, but eventually they may have evolved towards rearing younger siblings instead of their own children.

In the termites there may have been an additional ecological pressure for staying at home. These insects digest cellulose by means of protozoans living in their intestine, and the protozoa have to be passed literally from mouth to mouth between generations. Thus ancestral young termites had to stay at home long enough to become infected with cellulose digesting protozoa. As a thought provoking aside it is worth mentioning Richard Dawkins' (1979) suggestion that termite eusociality evolved because the protozoans manipulate termites into staying at home to build an ideal environment for protozoans to grow and replicate!

(b) Genetic predispositions

By helping to rear siblings, the workers pass on their genes to the next generation, so there is an obvious genetic predisposition towards helping. But Ric Charnov (1978) has pointed out that there is more to it than this: a mother makes a genetic gain by 'persuading' her children to stay at home and rear younger siblings while the children themselves make no genetic loss and therefore are 'willing' victims of maternal persuasion. The argument which is a development of Alexander's hypothesis of 'parental manipulation' (Alexander 1974) goes like this. Suppose there is enough time to rear two broods in a season and the queen could either rear the first brood, die, and let her daughters lay eggs and rear the second brood, or she could lay both lots of eggs and persuade her first batch of daughters to stay at home

195

*Chapter 10
Co-operation
and altruism
in the social
insects*

and rear the second brood. In the first case, the young emerging at the end of the season will be grandchildren of the original queen, related to her by 0.25 (chapter 1). In the second case they will be children of the queen, related to her by the usual 0.5. Therefore if the numbers of offspring at the end of the season are the same in the two cases, the queen will have doubled her genetic representation by persuading her daughters to stay at home and help. If the mother has a very large capacity to lay eggs, it is possible that she could supply each of her daughters with as many eggs as they themselves could lay and rear.

There is therefore strong selection on the queen to 'persuade' her daughters to help. But, and here we come to the crucial bit of the argument, the daughters lose nothing by staying at home. They rear younger siblings ($r = 0.5$) instead of children ($r = 0.5$) (assuming the original queen mated once at the beginning of the season). As long as the queen can lay enough eggs, the daughters make the same genetic gain from rearing siblings as from rearing offspring (Fig. 10.3).

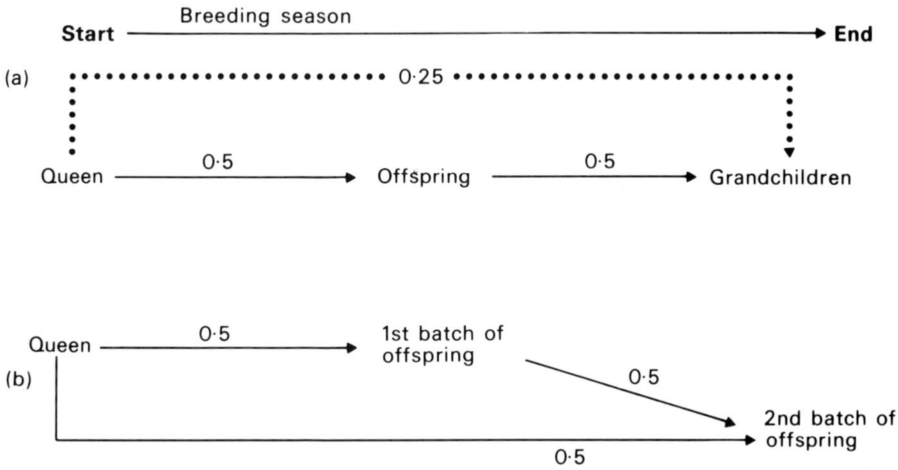

Fig. 10.3 An illustration of Charnov's (1978) idea that offspring may loose nothing by staying at home to help their mother, while the mother makes a large genetic gain. In (a) the queen starts the season by having children and ends with grandchildren (related by 0.25). In (b) she has two successive batches of children and ends with offspring related by 0.5. In (a) the offspring care for their own children and in (b) for their younger siblings, both related by 0.5.

HYPOTHESIS TWO: SHARING A NEST

In many tropical and some temperate zone wasps, nests are founded not by a single queen but by a group of co-operating individuals. Very often the foundresses are sisters (e.g. *Trigonopsis cameroni*) but sometimes they are not sisters and may possibly be unrelated (e.g. *Cereris hortivaga*) (West Eberhard 1978a). In primitively social wasps each queen lays her own eggs and rears her own young, but if one queen succeeded in dominating the others and preventing their eggs from developing, the stage would be set for the evolution of workers.

196

*Chapter 10
Co-operation
and altruism
in the social
insects*

(a) Ecological constraints

The ecological pressures favouring nest sharing are probably similar to those suggested earlier as favouring young that stay at home to help; defence against parasites and nest construction. For example in the social wasp *Trigonopsis cameroni*, 1–4 sisters build a communal nest out of mud. Although each female builds her own brood cell, lays her own eggs, and provisions her own young, the sisters co-operate in building shared walls of the nest and in driving away ants or other enemies (Eberhard 1972). The way in which this kind of co-operation between sisters might lead to an unequal division of reproductive labour is illustrated by *Metapolybia aztecoides*, another neotropical wasp, this time with a sterile worker caste. Here a nest is built by several queens, all of which lay eggs which develop into workers, that in turn help to enlarge the nest. Co-operation between queens is essential at the start in order to produce enough workers to get the colony and nest going, but once the colony is well established, one queen fights with and evicts the rest (that is if more than one queen is still surviving) before they have a chance to produce any reproductive offspring. Each queen therefore starts the colony with a certain chance of ending up having reproductive offspring, and the losers are 'hopeful reproductives' whose only chance at the start was to co-operate with the other queens. In *Metapolybia* the foundresses are probably sisters, so even the losers make a genetic gain from the colony, but if the chances of reproduction without co-operation were small enough, one could imagine co-operation even between unrelated foundresses (West Eberhard 1978b).

Nest sharing might also arise by accident. In the great golden digger wasp (*Sphex ichneumoneus*) two unrelated females sometimes end up using the same burrow because the second female was attempting to take over an apparently abandoned nest hole which happened to be occupied (chapter 8). Although in *Sphex* it is disadvantageous for females to share a burrow because they fight with each other and steal each other's prey, it is possible that accidental sharing could be the start of communal nesting if ecological pressures from parasites, for example, were very great.

(b) Genetic predispositions

If, as in *Metapolybia*, the foundresses are sisters, even those that fail to reproduce will still gain some genetic representation in the next generation. This is nicely illustrated by Bob Metcalf's study of paper wasps (*Polistes metricus*) (Metcalf & Whitt 1977). These wasps sometimes found nests as solitary queens and sometimes two sisters share a nest. When two queens share, one (the α female) does nearly all the egg laying while the other (the β female) passes on her genes largely through her sister's offspring. Metcalf calculated the degree of

197

*Chapter 10
Co-operation
and altruism
in the social
insects*

relatedness between sisters sharing a nest (by means of electrophoreti-
cally detectable enzyme polymorphisms) and also estimated the
number of offspring produced by α, β and solitary females. These
calculations showed: that the α female does better than the β, while
the β does about as well as a solitary queen in terms of genetic
representation in the next generation, and that the β female's genetic
contribution is made almost entirely through nephews and nieces.
Nests of pairs of females produce more young than those of solitary
females because they are better guarded against predators and parasites. Metcalf's study shows, therefore, that a *Polistes* female can do as
well by joining her sister as by setting up a nest on her own, even if she
produces hardly any children in the former case (Table 10.1).

Table 10.1 Metcalf's calculations of relative inclusive fitness in *Polistes metricus*.
The table shows a comparison between the gene contributions of solitary queens and
α and β queens that share a nest. The shared nests have a higher probability of
success than solitary nests. α females produce most of the young in a shared nest,
while β females make most of their gene contribution through helping their sisters'
young. The measures of success and fitness are expressed relative to that of a solitary
queen:

	Solitary queens	Joint nesters	
		α	β
Relative probability of nest success ($\pm$ S.E.)	1	1.38 ± 0.02	1.38 ± 0.02
Average relatedness to offspring	0.47	0.45	0.34
Relative inclusive fitness	1	1.83 ± 0.57	1.39 ± 0.44

The fact that social insect colonies are often founded by sisters
raises the interesting question of how siblings recognise each other. If
dispersal is very limited a simple rule such as 'co-operate with the first
queen you meet' might be adquate, but in at least some species there is
a genetically based ability to recognise relatives. Greenberg (1979) has
shown that workers of the sweat bee *Lasioglossum zephyrum* selectively prevent unrelated bees from entering their nest. There is a linear
correlation between degree of relatedness and the tendency of workers
to allow intruders into the colony. Greenberg suggests that the
recognition is based on genetically determined family odours and
discrimination by workers against bees with an unfamiliar smell.
    To summarise the discussion of how sterile castes might have
originated in the social insects, the two main hypotheses are that they
arose by offspring staying at home to help their mother, or by sisters
(and perhaps unrelated queens) sharing a communal nest. In both
cases the ecological pressures favouring sociality were probably nest
defence and nest construction, and the genetic predisposition towards
helping was that the recipients of aid were close relatives of the
altruists. The two hypotheses are not mutually exclusive, and most

198

Chapter 10
Co-operation
and altruism
in the social
insects

authors agree that the first is probably correct for wasps, termites, ants and some bees, while the second applies to the remaining eusocial bees. The two hypotheses are equally applicable to any sexually reproducing diploid organism, but we now turn to a very important special feature of the Hymenoptera which gives them an additional predisposition towards eusociality.

Haplodiploidy and altruism

W. D. Hamilton (1964) was the first to fully appreciate the significance of the special genetic predisposition of Hymenoptera to form sterile castes. The special feature is *haplodiploidy*: males develop from unfertilised eggs and are haploid, while females develop from normally fertilised eggs and are therefore diploid.

A haploid male forms gametes without meiosis, so that every one of his sperms is genetically identical. This means that each of his daughters receives an identical set of genes to make up half her total diploid genome. With a diploid father, a female would stand a 50 per cent chance of sharing any particular one of his genes with her sisters, but with a haploid father she is certain to share all of them. The other half of a female hymenopteran's genes come from her diploid mother, so she has a 50 per cent chance of sharing one of her mother's genes with a sister (Fig. 10.4). If we now think about the total degree of relatedness between sisters we come to a remarkable conclusion. Half

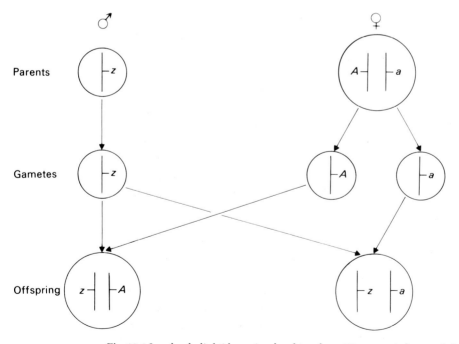

Fig. 10.4 In a haplodiploid species daughters have 50 per cent chance of sharing a gene from the mother (*A* or *a*) and 100 per cent chance of sharing a gene from the father (z).

their genome is always identical, and the other half has a 50 per cent chance of being shared, so the total relatedness is $0.5 + (0.5 \times 0.5) = 0.75$. In other words, because of haplodiploidy full sisters are more closely related to one another than are parents and offspring in a normal diploid species. Hymenopteran queens are diploid and are therefore related to their sons and daughters by the usual 0.5 (Table 10.2). A sterile female worker can therefore make a greater genetic profit by rearing a reproductive sister than she could if she suddenly became fertile and produced a daughter! This extraordinary state of affairs also suggests why in the Hymenoptera only females help to rear sisters. Males are related to their sisters by 0.5 instead of the 0.75 value calculated above for females. The relatedness of males to their sisters is calculated as follows: a haploid male inherits all his genes from his mother and there is a 50 per cent chance that a given sister will have inherited the same maternal genes. (Note, however that a female is related to her brother by only 0.25, since the 50 per cent of her genes that come from her father have no chance of being shared with a brother, and the other half of her genes have a 50 per cent chance of being shared: $0.5 \times 0.5 = 0.25$.)

199
Chapter 10
Co-operation
and altruism
in the social
insects

Table 10.2 *Degrees of relatedness between close relatives in a haplodiploid species*

	Mother	Father	Sister	Brother	Son	Daughter	Niece or nephew (via sister)
Female	0.5	0.5	0.75	0.25	0.5	0.5	0.375
Male	1	0	0.5	0.5	0	1	0.25

By contrast, in the termites in which both males and females are equally related to their siblings, both sexes are sterile.

Haplodiploidy might also help to explain why sterile castes have evolved more often in the Hymenoptera than elsewhere in the insects: Wilson (1975) gives a figure of 11 independent evolutionary origins for sterile castes in the social ants, bees and wasps, which constitute only 6 per cent of all insect species, and only one origin of sterility in all the rest of the insects, namely the termites. Since Wilson wrote this, non-reproductive soldiers have been discovered in a Japanese aphid (Aoki 1977) but aphids are genetically even more prone to sterility than the Hymenoptera. They reproduce asexually (for at least part of the year), which means that the members of an aphid colony are genetically identical just like the cells of a body. So the existence of sterile workers in aphids is perhaps no more of an evolutionary suprise than the fact that our nose cells do not produce sperms or eggs.

Although hapodiploidy in the Hymenoptera clearly predisposes them to become eusocial, it does not *cause* eusociality to evolve. This is easy to see because not all haplodiploid insects have sterile castes, and in the termites steriles castes have evolved in a normal diploid species. As we have stressed earlier in the chapter, ecological pressures and genetic predispositions work together to determine

200

Chapter 10
Co-operation
and altruism
in the social
insects

whether or not sterility will evolve. Haplodiploidy paves the way for the evolution of sterile castes because the critical ratio of benefits to costs (chapter 1) which has to be exceeded for helping sisters to be favoured is 4:3 instead of 2:1 as in normal diploid siblings.

A secondary cautionary note is that the simple calculations of relatedness in table 10.2 hold only if the colony is formed by a single queen who has mated once. If the queen mates twice and sperm mixes at random the average relatedness between sisters is only 0.5. If a colony is founded by several sisters as in *Trigonum*, helping is still more likely to evolve in a haplodiploid than in a diploid species because in the former a female is related to her sister's children by 0.375 and in the latter by 0.25.

Conflict between workers and queen

Our discussion so far has referred primarily to evolutionary history; we have described the ecological pressures and genetic predispositions that might have been important in leading to the evolution of sterile castes. In this section we are not going to discuss the origin of eusociality, but the selection pressures that act within present-day hymenopteran colonies. Our question will be: 'Given that there are sterile castes, how do workers and queens maximise their genetic profit?'. The answer to this question will probably tell us something about the selective forces *maintaining* eusociality today, but can be used only as indirect evidence for its origin.

Hamilton's theory, which we described in the last section, can be used to analyse how workers and queens might maximise their genetic profit. As we shall show, the theory predicts a conflict of interest between workers and queen over the sex ratio of reproductives in the colony.

CONFLICT OVER THE SEX RATIO

Our account of Hamilton's theory can be summarised as follows. Imagine a young female with the hypothetical choice of going off to rear her own daughters or staying at home to rear a new generation of younger reproductive sisters. Since she is more closely related to sisters than daughters, she would do better to stay at home and have sisters rather than the same number of daughters. In fact the queen seems to be the loser since she is condemned to have offspring!

However there is a further twist to the story that we have not considered. It may well be better for a young female to stay at home and rear sisters but this presupposes that the queen is going to produce sisters for her to rear. Obviously what the daughters can do will depend on what the queen is doing. As we saw earlier in the book, whenever we encounter situations like this we need to analyse the problem in terms of what will be the stable strategy or ESS.

201
Chapter 10
Co-operation
and altruism
in the social
insects

Let us consider the queen first of all. She is equally related to her sons and daughters ($r = 0.5$ in each case) and so just like diploid females of any sexual species she is expected to produce equal numbers of male and female reproductive offspring. To be more precise she should *invest equally* in the two sexes (chapter 6). It is important to emphasise that we are referring to equal investment in *reproductive* offspring, not sterile workers. Recall that in chapter 6 the argument was that a 50:50 sex ratio was stable because the expected *reproductive success* of a male and a female is the same. Hence the discussion of sex ratios is only pertinent to reproductives.

Now for the twist; if the queen produces an equal sex ratio, workers will spend their lives rearing equal numbers of brothers (to which they are related by 0.25) and sisters (related by 0.75). Their average relatedness to reproductive siblings will therefore be only 0.5 exactly the same as they would have to their own progeny if they had decided to leave home and have their own offspring!

In order for female workers to gain the full genetic benefit from staying at home and rearing sisters they must rear more queens than drones. But how much bias in favour of reproductive sisters should they show? Once again we search for the ESS sex ratio, this time from the workers' point of view. The workers are more closely related to their sisters and so should rear more of them than brothers. But if they rear too many sisters then the sex ratio in the population will become so female-biassed that a drone will have very much greater reproductive success than a queen. The stable sex ratio for the workers is 3:1 in favour of reproductive females. When female reproductives are exactly three times as common as males, drones have three times the expected success of queens because on average each drone can mate with three females. From the workers' point of view this would exactly compensate for the fact that brothers are only one third as closely related as are sisters: a worker expects to get three nieces or nephews from her brothers for every one she gets from her sisters. Nieces and nephews on her sister's side are three times as closely related to her, so the total gain per unit investment via brothers and sisters is the same.

To summarise this rather complicated argument, the queen prefers to an equal investment in male and female reproductive offspring, but the workers prefer a ratio biassed 3:1 in favour of females. There is a direct conflict of interest over the sex ratio between workers and the queen. Who wins?

Tests of worker-queen conflict

Bob Trivers and Hope Hare (1976) attempted to test whether the queen or workers win by analysing the ratio of investment (more accurate than simply looking at numbers) in male and female offspring in 21 species of ants. The ant species were chosen because they were ones in which the conditions for the hypothesis apparently hold (one queen,

202

Chapter 10
Co-operation
and altruism
in the social
insects

one mating). Despite a considerable amount of scatter in their data, Trivers and Hare found that on average, the ratio of investment was much closer to 3:1 than to 1:1 (Fig. 10.5). They concluded that the workers win the conflict and successfully manipulate the sex ratio towards their own optimum and away from that of the queen. To put it baldly, the workers are successfully farming the queen as a producer of nieces and nephews: a far cry from the idea of workers as subordinate females making the best of a bad job! Trivers and Hare suggest that the workers win simply because they have practical power; they provision the young and are in a position to selectively kill off males and nurture queens. The queen presumably retaliates during evolution by attempting to control worker behaviour with pheromones or direct aggression.

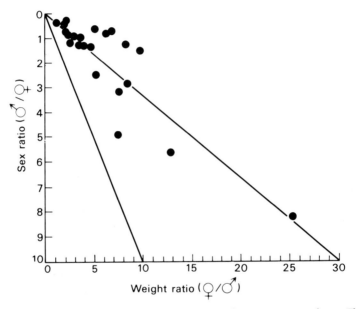

Fig. 10.5 Ratio of investment (measured by weight) in 21 species of ants. The x-axis is the ratio of female:male weight and the y-axis is the ratio of numbers of males:females in the colony. The lower line is the prediction if the investment ratio is 1:1 and the upper line is 3:1 in favour of females. The data are closer to the 3:1 line, as predicted if workers control the sex ratio. (To understand how the lines are drawn take the example of a ♀:♂ weight ratio of 6:1. Equal investment would mean six ♂ per ♀, and a 3:1 investment ratio in favour of ♀ would mean a ratio of 2 ♂ per ♀). (Trivers and Hare 1976).

Although Trivers and Hare considered two extreme divisions of practical power, total queen control and total worker control of the sex ratio, a more plausible alternative is that power is shared. The queen can choose the sex of the eggs she lays (by whether or not she fertilises them with stored sperm) and the workers can choose whether or not to rear the larvae. With shared power the problem becomes 'given that the workers control the sex ratio after egg laying, what ratio should the

203
Chapter 10
Co-operation
and altruism in
the social
insects

queen lay', and 'given that the queen lays a certain ratio of eggs, how should the workers manipulate it'. These questions still remain to be answered, but they may yield more refined predictions than the simple ones tested so far.

Alexander and Sherman (1977) have criticised Trivers' and Hare's paper on four grounds.

1   They question whether the assumption that the queen mates only once in the ant species involved is really correct.

2   They emphasise that workers sometimes lay unfertilised ($\male$) eggs and that therefore the 3:1 sex ratio predicted by Trivers and Hare may be incorrect. If for example all the males were her sons, the ratio of relatedness (and therefore preferred sex ratio) for a worker would be 3:2 instead of 3:1.

3   They point out that there is a very large amount of unexplained scatter in the data.

4   Most important of all, they offer an alternative explanation for the female biassed sex ratio. In chapter 6 we mentioned that Fisher's theory of equal investment in males and females no longer holds when there is competition between brothers for mates (local mate competition). If a mother 'knows' that her daughters will all be fertilised by her sons, she should produce a brood made up mostly of daughters, with just enough sons to do the fertilising. Alexander and Sherman suggest that there may be some local mate competition which accounts for the biassed sex ratio in the ants studied by Trivers and Hare. This hypothesis is not incompatible with the idea of worker manipulation. If the queen favours a biassed sex ratio because of the effects of local mate competition the workers will favour an even more biassed ratio because of the additional impact of 0.75 relatedness. Without knowing the exact degree of local mate competition it is not possible for Alexander and Sherman to make a quantitative prediction of the optimal sex ratio for the queen, and the 3:1 bias observed by Trivers and Hare must be taken as suggestive, but not conclusive, evidence for the idea of worker manipulation.

It will be apparent from the discussion of Trivers and Hare's paper that in order to test whether workers or queens control the investment ratio we need to know about the extent of local mate competition, whether or not workers lay eggs, and the degree of relatedness of workers and reproductives. Bob Metcalf has succeeded in collecting all this information and measuring the ratios of investment in two species of *Polistes* wasp. The two species, *P. metricus* (referred to earlier) and *P. apachus* (Fig. 10.6) are both annuals. Their life histories are as follows. Nests are founded by one or more mated sisters in the spring. The nests produce reproductive males relatively early in the summer, followed by queens at a later stage; workers are produced continually through the season. The young queens mate in the late summer and over winter before starting up their own nests the following spring.

204

Chapter 10
Co-operation
and altruism
in the social
insects

Fig. 10.6 *Polistes apachus* workers on the nest. This species has a female biassed sex ratio of reproductives, perhaps indicating that workers benefit more than the queen (Photograph by permission of Bob Metcalf).

Metcalf's results were very striking. In *P. metricus* the ratio of investment was 1:1 (Metcalf 1980), while in *P. apachus* it was 2.48:1 (Metcalf and Finer 1981). Genetic analyses of enzyme polymorphisms showed that workers are related to their reproductive sisters by an average of 0.65 (slightly less than the theoretical maximum of 0.75 because the queen sometimes mates more than once). Also, workers do not normally lay eggs. There was little evidence of local mate competition: this was inferred from the fact that there is little or no inbreeding and the observation that females do not disperse far. Males must therefore disperse rather a long way and so brothers are not likely to compete. The conclusion seems to be that in both species workers should prefer a female biassed and the queen a 1:1 ratio of investment in reproductives.

205
Chapter 10
Co-operation
and altruism
in the social
insects

First consider the species found to have a 1:1 ratio. If there is worker–queen conflict as predicted by Hamilton's theory, the queens appear to win. Metcalf suggest that they win because of practical considerations. How might the workers manipulate the sex ratio? The obvious way would be by killing off male eggs or larvae (given that the queen lays eggs in a 1:1 ratio and that workers cannot lay female eggs themselves this seems to be the only obvious possibility). In *P. metricus* the queen produces male offspring early in the season. At this stage few of the workers have emerged and the few that have emerged can be effectively controlled by the queen. By this ploy it appears that queens have removed the opportunity for workers to control the sex ratio. The workers lose, in that they rear reproductives with an average degree of relatedness of 0.45 (0.25 for brothers and 0.65 for sisters) while the queen is related to her offspring by 0.5.

What about *P. apachus*? If there is worker–queen conflict in this species, the workers appear to succeed in manipulating the sex ratio in their favour. Metcalf argues that the crucial difference between the species lies in the length of their breeding seasons. In *metricus* it is May–October, while in *apachus* it is March–November. The extra length of the breeding season means that in their evolutionary past *apachus* workers had an option open to them which was not open to *metricus* workers, namely the option to leave home and start a nest. (This is clearly not an option open to present day sterile workers. What we are referring to are hypothetical alternative evolutionary strategies which may have been open in the past). The breeding season of *metricus* would have been too short for workers to leave after males have matured, mate, and set up their own nest, while the extra three months of the *apachus* season means that this would have been a feasible option. Only if *apachus* workers could succeed in biassing the sex ratio sufficiently would 'stay at home' be more advantageous than 'leave'. The observed bias (71 per cent females) means that the average relatedness between a worker and the reproductives it rears is $(0.71 \times 0.65) + (0.29 \times 0.25) = 0.53$, which is better than the relatedness of 0.5 between a mother and her offspring. If all other things were equal 'stay at home' would be a better policy than 'leave'. Metcalf argues, therefore that sterile helpers have persisted in *apachus* because workers have manipulated the investment ratio to their advantage.

The interpretation of these results is still speculative but the general messages are as follows: the outcome of worker-queen conflict may vary from species to species, and whether queens or workers win depends critically on practical considerations which determine the options open to either party.

206

Chapter 10

Co-operation

and altruism

in the social

insects

The evolutionary battle between workers and queen over the sex ratio has taken place in spite of the fact that queen and workers are not genetically differentiated. It is the diet and not the genes of a female that determine whether she becomes sterile or reproductive. If one thought in terms of genes for controlling the sex ratio, these genes must produce a conditional strategy 'if in a worker body favour 3:1, if in a queen favour 1:1'.

Comparison of vertebrates and insects

There are no known examples of sterile castes in vertebrates, but in other respects there are some close parallels between the conclusions of this chapter and those of chapter 9. The two hypotheses about the evolutionary origin of helping in insects are similar to the two kinds of co-operative nesting described for birds in chapter 9. The first resembles scrub jays, in which young birds stay at home to help their parents rear younger siblings, while the second is parallel to anis and ostriches where several females share the same nest and one female is dominant over the rest and produces more young.

The ecological pressures thought to favour helping in vertebrates and insects are also quite similar. For example both scrub jay helpers and workers in social insects play a similar role in defence of the nest against predators or parasites. Nest sharing by reproductive females is also thought to be a response to predator pressure in both vertebrates and insects. Another major ecological factor which leads to helping in birds is lack of opportunity for young birds to find a nesting territory and this again is paralleled by the importance of co-operative nest building in social insects: a lone female has little chance of building a nest alone, so she is forced to co-operate.

In spite of these similarities there must also be important differences which have lead insects but not vertebrates to evolve sterile castes. The differences could either by ecological or genetic. First, the ecological and life history constraints are different in insects and vertebrates.

Vertebrate helpers such as young scrub jays help as part of a long term strategy of acquiring a breeding territory or mate in the future. In short-lived insects, such long term gains are less (although there are exceptions such as *Metapolybia*), and the emphasis is shifted to gains from helping siblings. If a young scrub jay, for example, stays at home to help, it does less well than one that succeeds in setting up on its own (chapter 9). This is because the mother's brood does not provide enough young for the helper to be able to rear as many as it could produce if it bred on its own, the constraint on the mother being her inability to incubate a very large clutch. An insect queen, in contrast, can often provide her daughters with as many eggs as they themselves could lay, because the constraint on brood size is the availability of nest chambers and not the need for incubation. For both insects and

207
Chapter 10
Co-operation
and altruism
in the social
insects

vertebrates one constraint which prevents helpers from setting up on their own is the problem of obtaining a nest or territory. In insects nest construction takes relatively longer than in vertebrates, so this constraint is more severe. In other words, a nest is a scarcer resource for an insect than is a territory for a vertebrate.

The differences between birds and insects in genetic predispositions towards sterility have already been dealt with in depth. Hymenopteran haplodiploidy and aphid asexuality are features that are not shared by any of the vertebrates discussed in chapter 9. The termites appear to have no particular genetic predispositions towards sterility and are therefore more closely parallel to vertebrates. Perhaps an important factor here is the sedentary habits of termites. Because they do not disperse far, individuals in a colony are closely inbred and therefore closely related.

Summary

In the social insects there are sterile workers that never have offspring but instead help to rear younger siblings. This appears at first sight to go against the idea of natural selection favouring maximum efficiency at passing on genes. However the fact that sterile workers rear close relatives genetically predisposes them to be altruistic.

There are two theories about how sterile helpers arose in evolution: by young staying at home to help with mother and by sisters sharing a nest with one producing most of the offspring. According to both theories the ecological advantages of communal nesting are defence against parasites or predators and construction of an elaborate nest. The genetic prediposition is that help is given to close relatives. In the case of the first theory there may be strong selection for mothers to persuade their daughters to stay at home and little selection for daughters to resist.

In the Hymenoptera there is an additional genetic predisposition towards helping: haplodiploidy. Sterile female helpers in the Hymenoptera could pass on their genes even more efficiently than the reproductive queen if they bias the sex ratio of the reproductive siblings they help to rear. This is because sisters are more closely related to one another than are mothers and offspring. There is suggestive, evidence for a female bias in the sex ratio in some species.

Further reading

Hamilton (1972) summarises the importance of haplodiploidy in Hymenoptera as a predisposition towards eusociality.

Evans (1977) describes how eusociality might have evolved because ecological constraints favoured offspring helping their mothers.

West Eberhard (1978b) critically reviews ideas about the origin of eusociality and favours the hypothesis that it arose from associations between foundress sisters.

Wilson (1975) in the chapter on social insects gives an excellent review of the biology of this group.

Michener (1974) is the definitive work on social bees.

Dawkins (1979) enumerates an entertaining and salutory list of twelve common misunderstandings in the literature on kin selection. For example the statement that 'all individuals in a species share a high proportion of their genes (shown by DNA hybridisation studies) and they should therefore be altruistic to one another' is fallacious. To see why, look at Dawkins' paper.

Wilson (1980) discusses an aspect of social insect behaviour not covered in this chapter, namely the optimal allocation of workers in a colony to different tasks. He takes the example of leaf cutting ants (*Atta*) and shows that one size of worker is more efficient than the others at cutting and carrying leaves for the colony's store. This is the size at which workers normally change from nest maintenance to foraging duties. A general introduction to the economics of caste in social insects is given by Oster & Wilson (1978).

Chapter 11. The design of signals: ecology and evolution

Most of the interactions between individuals described in this book involve communication. Males attract females or repel rivals, offspring beg from their parents, and poisonous caterpillars warn their predators by using special *signals* or *displays*—behaviours and structures designed for use in communication. This chapter is about how signals are designed by natural selection for effective communication. We will discuss the influence of two kinds of selection pressure on signals: ecological constraints imposed by the environment and the response of reactors at whom the signals are directed. But first let us clarify what we mean by communication.

The most obvious characteristic is usually that a signal or display of one individual (actor) in some way modifies the behaviour of another (reactor). The reactor's response may be immediate and obvious (a male firefly rapidly flies towards a flashing female of the correct species), it may be subtle and difficult to detect (a male antelope slightly alters its direction of walking to avoid crossing a territorial boundary when it detects the scent marks of a resident), it may be delayed in time (the ovaries of a female budgerigar gradually develop as the result of the stimulus of male song), or it may not occur all the time (a territorial male blackbird sings for several hours during which time only one or two intruders hear the song and retreat). If, in spite of these difficulties, we can detect the response of a reactor, we could characterise communication as the *process in which actors use specially designed signals or displays to modify the behaviour of reactors*. The qualification 'specially designed signals' saves our definition of communication from becoming too wide. When we see a drunkard lurching down the street towards us late at night we may cross the road to get out of the way, but since there is no reason to suspect that the drunkard's lurch has been developed by natural selection to signify advanced inebriation we would not include lurching within the definition of communication. In contrast, the grasshopper's incessant buzz, which is created by rubbing the legs together and probably evolved from simple walking movements, is an example of communication. The ancestral male grasshopper may have lurched towards a female accidentally producing a chirrup at the same time, but natural selection has acted on the sound to make it into a loud and conspicuous signal for attracting females.

As an aside it is worth pointing out that our definition of communication would not satisfy all students of behaviour. Our concern in this chapter is with how signals evolve, and we therefore stress the importance of specially adapted signals in communication. But for someone interested in human or animal social interaction a broader definition 'any aspect of A's presence or behaviour which influences B' might be appropriate. This would encompass observations such as those of Michael Argyle (1972) who found that many subtle aspects of posture (leaning back in a chair, crossing legs etc.) play a major role in human communication, even though they may have not evolved as special signals.

Ecological constraints and communication

Different groups of animals rely on different sensory channels for communication. Small mammals live in a world of smells, birds in a world of music, and coral reef fish in a world of brilliant poster colours. Why are there such differences? Part of the answer is that the utility of different channels depends on constraints imposed by the habits and habitats of a species. It is obvious, for example, that sound or scent are more useful than visual signals for nocturnal small mammals and that birds living in dense bushes can more readily hear than see each other. Roe deer living in dense forest mark their territories with loud calls and scents deposited on the vegetation, while in open habitats they use primarily visual signals (Loudon pers. comm.). But differences in the effectiveness of transmission are not the only considerations in assessing the costs and benefits of various communication channels (Table 11.1). Sound is very flexible: enormous numbers of signals can be fitted into a short space of time by rapid changes in pitch, loudness and harmonic structure. Scent may be less flexible but is energetically cheap to produce and can last for a very long time, an advantage for an animal such as a fox with a very large territory in which it can announce its presence at any particular site only once every few hours, or even every few days. Brilliant body colours are permanently on display (at least seasonally); while this may be advantageous for attracting females and deterring rivals, it may be a considerable disadvantage in attracting unwelcome predators!

Table 11.1 Advantages of different sensory channels of communication (from Alcock 1979).

Feature of channel	Type of signal			
	Chemical	Auditory	Visual	Tactile
Range	Long	Long	Medium	Short
Rate of change of signal	Slow	Fast	Fast	Fast
Ability to go past obstacles	Good	Good	Poor	Poor
Locatability	Variable	Medium	High	High
Energetic cost	Low	High	Low	Low

The use of different communication channels in different ecological conditions is nicely illustrated by Bert Holldöbler's (1977) study of recruitment signals in ants. When ant workers return from a foraging trip to the colony they often recruit others to take part in harvesting the supply of food. Holldöbler describes three kinds of recruitment, two of which are illustrated in Fig. 11.1.

(a)

(b)

Fig. 11.1 Two types of recruitment communication in ants: (a) Tactile (b) Chemical.

1 Species such as *Leptothorax* feed on single, immobile prey items (e.g. dead beetles) which are too big for one worker to carry but which can be brought back to the nest by two individuals pulling together. After finding a prey, *Leptothorax* workers return to the nest, regurgitate some of the food and secrete a chemical signal from their abdomen to attract other workers. One worker is recruited to help bring back the prey and it is 'led by hand' to the foraging site—so-called 'tandem running' in which the recruit follows the leader by keeping contact with its body: the recruit's antennae rest on the leader's abdomen (Fig. 11.1a).
2 Fire ants (*Solenopsis*) feed on large mobile prey (large insects and so on) which need several individuals to carry them back to the nest and they use odour trails as a means of recruitment. After finding a suitable prey, a worker returns to the nest laying a trail of scent secreted from a special abdominal gland. The odour trail excites other workers to join in, running along the scent trail to the prey, and returning to the colony laying their own trail if they have found the prey (Fig. 11.1b). The scent trail builds up very rapidly as large numbers of workers add to it, but it also decays rapidly as soon as workers stop renewing it because the scent is very volatile and lasts only for a few minutes. This means that the trail persists only as long as the prey is available, and it can change position to keep up with the moving prey.

3 A third kind of ant foraging trail is characteristic of leaf cutters (*Atta*) and seed eaters such as *Pogonomyrmex*. These ants feed on long lasting or renewing patches of food, so that workers use the same trails for days, weeks or years on end. There are two ways of marking trails, by long lasting scents and by cutting a path through the vegetation. Both of these provide durable signals.

These examples illustrate how different kinds of signals using different sensory channels are used according to the feeding ecology of the ants: tactile signals for recruiting a single companion to stationary food, rapidly decaying odour signals for recruiting large numbers of workers to mobile prey, and visual cues or long lasting scents for large patches of renewing food.

BIRD AND PRIMATE CALLS

The way in which habitat structure and meteorological conditions affect signal transmission has been studied in detail for sound signals, especially bird songs. Differences between species and between populations within a species can sometimes be explained in these terms.

Gene Morton (1975) and Claude Chappuis (1971) were the first to show that the structure of bird songs is correlated with habitat. Morton found that the songs of species living below the canopy in tropical forests in Panama were characterised by lower frequencies, a larger proportion of pure tones and a narrower range of frequencies than songs of grassland birds in the same country (Table 11.2). Typically the songs of tropical forest birds contain many low pitched pure whistles and those of grassland species sound like buzzy trills.

Table 11.2 Differences between the songs of forest and grassland birds in Panama (Morton 1975).

Habitat	Emphasised Frequency (kHz)	Percent pure tones	frequency range (kHz)
Forest *closed* (Below canopy)	2.2 ↓	87 ↗	1.5 ↔ *narrow*
Grassland *open*	4.4 ↑	33 ↓	3.5 *wide*

Several subsequent studies have shown that geographical variation within a species can sometimes be correlated with habitat type. For example Fernando Nottebohm (1975) found the rufous-collared sparrows (*Zonotrichia capensis*) living in South American forests sing slower trills than their grassland cousins (Fig. 11.2a). The great tit, like the rufous-collared sparrow, lives in many different kinds of habitat over a wide geographical area. Its distibution stretches from Ireland to Japan and from Finnish birch forests to Malaysian mangrove swamps.

Mac Hunter and John Krebs (1979) recorded the territorial songs of great tits in two contrasting kinds of habitat, open woodland or parkland and dense forest. Regardless of geographical location (recordings were made in various countries stretching from Norway to

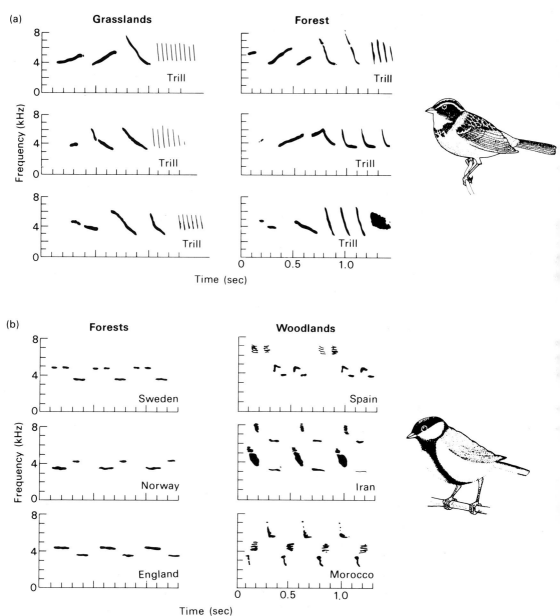

Fig. 11.2
(a) The song of the rufous-collared sparrow (chingolo) has slower trills in forests than in open country.
(b) Great tits in dense forests sing songs with a narrower range of frequencies, lower maximum frequency and fewer notes than the songs of open country birds.

Iran) there were consistent differences between songs from the two habitat types (Fig. 11.2b). The birds in more open habitats have songs with higher maximum frequency, more rapidly repeated notes, and a wider range of frequencies than those of forest birds. So striking are the correlations with habitat type that the songs of birds in a park in Southern England are more similar to those from the same sort of habitat in Iran (5000 km away) than they are to birds in a dense forest 100 km away in another part of Southern England.

What is the survival value of these differences in song structure between habitats? The answer is still not certain. Morton hypothesised that songs are designed to carry the maximum possible distance and suggested that the differences he observed between habitats might be related to differences in the attenuation of sound. He tested this by measuring the attenuation of pure tones of different frequency that he broadcast in the two habitat types from a loudspeaker. He found that in both habitats high frequencies attenuated more rapidly than lower ones. This is not surprising, since on theoretical grounds one would predict that high frequencies would be more readily attenuated by obstacles such as leaves and branches, by air turbulence, and by the viscosity of the air itself. The unexpected aspect of Morton's results was that sounds of about 2 kHz carried particularly well, better than higher or lower pitched tones, in forests but not grassland. Since forest birds sing at about 2 kHz, Morton suggested that their songs are tuned into the 'frequency window' of minimum attenuation. The reason for higher pitched, wider frequency band, songs in grassland was less clear.

However, more recent studies of sound attenuation have shown that there are probably only fairly small differences between habitat types (Marten & Marler 1977a, b; Wiley & Richards 1978). While in forests branches and leaves cause attenuation of high frequencies, in more open habitats local air turbulence often has a similar effect. Since open areas are often less sheltered than forests, the overall pattern of attenuation of different frequencies tends to be similar in the two habitats. Further, Morton's frequency window seems to be more related to attenuation of lower (below 2 kHz) frequencies by the ground than to differences between habitats.

Wiley and Richards (1978) propose an alternative explanation for the differences between forest and open habitats. They point out that there are two problems facing a singing bird. One is *attenuation* of sound: if the song attenuates too much the receiver will not be able to detect it because it is lost in the background noise. The other is *degradation*: if the song is degraded or distorted in its passage through the environment it may be confused by the receiver with other sounds. They suggest that there are much greater differences in degradation than in attenuation between habitats. In forests, the major source of degradation is echoes or reverberations from branches and leaves, while in open habitats the major source is irregular fluctuations in amplitude caused by gusts of wind that mask the song. These two kinds of degradation can be reduced by different design features. Reverberations are more severe for high frequency than low frequency sounds (because high frequencies are deflected by small objects such as leaves and branches). They also cause a problem if the song contains rapidly repeated notes as are found in the trills of many bird songs, because the echoes will become confused with the original notes. Therefore songs designed to overcome reverberations in forests should be of low frequency and contain either pure notes as opposed to trills, or trills with widely spaced notes. These are precisely the patterns observed by Morton and in the studies of chingolos and great tits. The irregular amplitude fluctuations of open habitats, in contrast, favour rapid trills. Since the song is masked by wind at irregular intervals it has to contain notes that are short enough and sufficiently rapidly repeated to be detected in a short space of time. The same need to fit the songs into short intervals between gusts may account for the high frequencies of open habitat songs, since a high frequency sound requires a shorter time for a given number of wavelengths. Again the patterns predicted by Wiley and Richard's hypothesis fit well with those observed in the field.

An additional complication is that in at least two of the studies referred to (Morton's and the great tit study) forest birds occupied larger territories than those in open habitats. The songs in the two habitats may be adapted not for maximum detection distance, but for an optimum distance which differs between habitats because of the difference in bird density. (An optimum distance could arise if, for example, it was disadvantageous to signal over too great a distance— see below.) Forests songs with their low frequencies and energy concentrated into a narrow band are likely to carry further than the higher pitched, wider frequency band songs of open country birds.

There is a striking parallel between these results of studies of bird song and the interpretation by Busnel and Klasse (1976) of a peculiar kind of human language. In four parts of the world, Andorra, Turkey, Mexico and the Canary Islands, local peasants have developed extraordinary whistling languages with elaborate vocabularies. Although the details of the languages and the methods of sound production vary from place to place, all are designed for long distance communication. The four places are all mountainous with steep sided, rocky valleys; distances are not very great as the crow flies but are large in terms of the effort needed to cross from one side of the valley to the other. The whistling languages are used for communicating across the steep valleys. They have all their sound energy concentrated into a narrow band of frequencies and are therefore well designed, like forest bird songs, to be detected and interpreted over relatively large distances.

Not all sounds are used for long distance signalling, and short distance sounds may be designed not to carry further than necessary.

The grey-cheeked mangabey (*Cercocebus albigena*) (Fig. 11.3) is a monkey which lives in the forests of east Africa; it lives in troops and defends a group territory. Peter and Mary Sue Waser (1977) studied two vocalisations of the mangabey, the 'whoop-gobble' which is used in intergroup signalling, and the 'scream' which is used during agonistic encounters within a group. The two sounds are produced at the same volume by the monkeys, but because the whoop-gobble is lower-pitched and has a narrower range of frequencies it carries much further than the scream (Table 11.3). The differences in frequency stucture and therefore carrying power of the two sounds reflect their design for communication over different distances.

Fig. 11.3 The grey-cheeked mangabey. (Photo by P. M. Waser).

Table 11.3 Comparison between two calls used by the grey-cheeked mangabey in tropical forests of Uganda (Waser & Waser 1977).

Call	Function	Distance from which heard by human ear	Sound pressure 5 m from monkey	Frequency (Hz)
Whoop-gobble	Spacing between troops	1000 m	75dB	300–400
Scream	Within troop agonistic encounters	300 m	78dB	1000–3000

Probably the reason why the mangabey uses a song which does not carry far for signalling within the troop is in order to avoid attracting predators. Another way to make sounds less likely to attract a predator is to produce unlocatable sounds, as was first suggested by Peter Marler (1955) in a classic paper about alarm calls of small birds. He pointed out that the calls given by many species of small birds when a hawk flies over are remarkably similar. All are thin high-pitched 'seeet' sounds which are hard for the human ear to locate. Marler suggested that the calls have evolved a structure which makes them hard for predators to locate and therefore reduces the risk to the caller of attracting the hawk to itself. (Why the caller should warn his flock mates in the first place is a different matter—see chapter 1.) Elegant and appealing though Marler's hypothesis is, an attempt to test its assumption that the calls are hard to locate did not support it. Michael Shalter (1978) tested pygmy owls (*Glaucidium* sp.) and goshawks (*Accipiter gentilis*) (both of which prey on small birds) to see if they could locate alarm calls. He put the predators on a perch between two loudspeakers and looked at their head movements to see if they could tell from which speaker the call was played. The birds seemed to be as good at locating alarm calls as they were with other supposedly more locatable sounds. The significance of the structure of alarm calls therefore remains something of a mystery.

CHOICE OF MICROHABITAT AND BROADCASTING TIME

(a) Microhabitat

Although the transmission of sounds is to some extent constrained by physical characteristics of the habitat, animals can increase the distance to which their calls carry by choosing an appropriate microhabitat and time of day for calling. For animals living and calling near the ground, the earth itself is a major cause of sound attenuation, and measurements show that attenuation diminishes markedly as a result of moving just a few feet up from the ground (Marten & Marler 1977). This might explain why so many song birds fly up from the ground to a prominent perch before starting to sing. The flightless burrowing cricket (*Anurogryllus arboreus*) usually lives on the ground but males climb up a tree before singing. By so doing they increase the area over which their song can be heard by fourteen fold (Paul & Walker 1979). Since the call is used for attracting female crickets, increasing the transmission distance must make an important contribution to reproductive success.

(b) Broadcasting time

> 'The busy lark, the messenger of day
> Sings salutation to the morning grey.'

Chaucer may have been one of the first English writers to comment on the dawn chorus, but field workers studying bird song and many other sounds know to their cost that dawn is the best time for collecting data. There are many environmental factors which contribute to the cost–benefit equation favouring dawn song, but an important one is daily variation in the extent to which sound is attenuated. As we have already mentioned, a major cause of sound attenuation is air turbulence, and in most habitats there is less turbulence at dawn or dusk than at other times of day—the familiar pattern is a tranquil period at dawn and dusk preceding and following a breezy day. This arises because local turbulence is largely caused by temperature gradients created by the sun. Henwood & Fabrick (1979) have calculated that a typical bird song will carry 20 times as far at dawn as at midday simply because of the differences in air turbulence affecting attenuation and the level of background noise. But this is probably not the only factor favouring dawn song. For example the benefit of doing other things such as feeding may be low at dawn, because the dim light makes it hard to find prey (Kalcelnik 1979a).

The cacaphony of dawn song is often so great that it is hard to pick out the sounds of any particular species. The same problem of acoustic interference probably confronts the singing animals themselves, and it may explain why different species sometimes occupy slightly different time niches in the dawn chorus. In the forests of Sumatra there are four species of calling primate, each of which sings at a slightly different time after dawn: first the orangutan, then the leaf-nosed monkey, the gibbon, and finally the siamang. In Borneo however, only the orangutan is found and its dawn chorus has expanded to fill the whole of the period occupied by four species in Sumatra (Mackinnon 1974).

To summarise our discussion of ecological constraints, sometimes the use of different sensory channels by different species can be interpreted in terms of ecological costs and benefits, as was illustrated by the ants. Within a sensory channel the exact structure of the signal as well as the time and place from which it is produced can be related to constraints on transmission imposed by the environment. This has been best studied in bird and monkey calls.

Reactors and the design of signals

The constraints of the environment impose broad limits on the design of signals, but within these limits the way in which signals evolve results from selection to increase their effectiveness in altering the behaviour of reactors. Reactors play an important role both in the evolutionary origin of signals and in their subsequent evolution towards increased effectiveness.

To the casual observer animal displays often seem to be inexplicably bizarre and absurd. Why should male ducks perform sham drinking and preening movements as part of their courtship ritual? Why do wolves mark their territories by urinating? Why do rhesus monkeys grin as a signal of fear and appeasement?

A great step forward in understanding the answers to these questions came when ethologists such as Lorenz and Tinbergen realised that many signals have evolved from incidental movements or responses of actors which happened to be informative to reactors. Selection favoured reactors who were able to anticipate the future behaviour of actors by responding to slight movements which predicted an important action to follow. If a dog always bares his teeth before biting, reactors who are able to anticipate and escape from an aggressive attack by observing bared teeth will be favoured by selection. Once this happens selection will favour actors who bare their teeth as a means of rapidly deterring reactors, and teeth baring will begin to evolve into a threat display.

We should expect then, that the incidental movements and responses from which signals have evolved are those which were originally most informative about future actions. This is borne out by many studies of the detailed form of displays in birds, fish and mammals. Many signals in these animals have apparently evolved from intention movements such as are made by a bird when it crouches and tenses its muscles for take-off: it is not hard to imagine that intention take-off movements were originally good predictors of a lunging attack by a rival, approach by a prospective mate and so on. Other movements performed at moments of transition from one major activity to another are also frequently the raw material from which displays have evolved. Often these movements reflect motivational conflict or indecision as the animal vacillates between, for example, attack and running away (Table 11.4).

Our examples of the duck, wolf and monkey can be interpreted in a similar way. The drinking and preening movements of courting ducks are probably derived from displacement activities—seemingly irrelevant actions which tend to occur at moments of balance between incompatible motivational states such as aggression and sex. Urinating is like blushing and hair erection, which are also the basis for displays in some species, a consequence of autonomic nervous activity during moments of stress. The ancestral wolf may have urinated uncontrollably when confronted with a rival at its territory boundary (anyone who has nursed a nervous dog will know all about this sort of thing!), and the response has subsequently evolved into a way of signalling 'keep out'. The grin of the terrified monkey is very similar to the reflex response with which a monkey protects the most vulnerable parts of his face such as eyes and mouth from the onslaught of an attacker; as with urinating, an autonomic reaction to stress has become a signal during evolution.

Table 11.4 Examples of the kinds of behaviour patterns and other responses from which displays in birds, fish and primates are thought to have evolved (Hinde 1970).

Behaviour or response	Example of display
1. Intention movement	Sky pointing in the gannet
2. Ambivalent behaviour	Oblique threat posture of black headed gull
3. Protective response	Primate facial expressions
4. Autonomic response (e.g. sweating, urinating, rapid breathing)	Vocalisations (from rapid breathing). Scent marking
5. Displacement activities	Preening in duck courtship
6. Redirected attack	Ground pecking in herring gulls

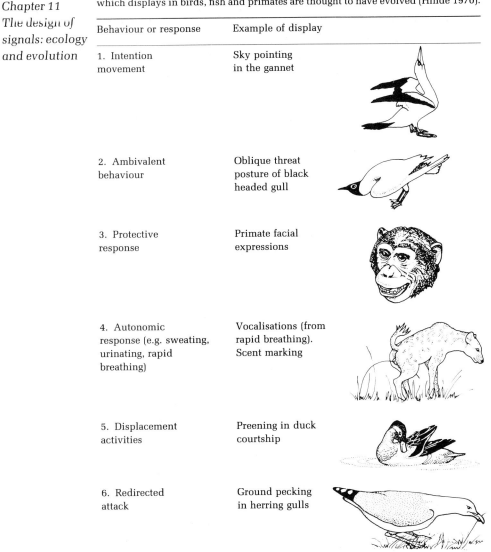

Most of the conclusions described in the above two paragraphs are inferences based on studies of the detailed structure of displays and the contexts in which they occur. For example the observation that threat displays occur at territory boundaries and in sequences of behaviour which include transitions between attack, threat and retreat suggests that threat signals arise when the animal is in motivational conflict. Similarly the structure of displays can be viewed as reflecting motivational conflict. The 'zig-zag' display of the three-spined stickleback (*Gasterosteus aculeatus*) involves a strange movement in which the courting male approaches the female in a series of short arcs, as if he is in a conflict between approach and avoidance.

There is also some more direct evidence for the conflict hypothesis from experiments in which the tendencies of an animal to attack and flee are manipulated independently. Nick Blurton Jones (1968) studied the threat displays of captive great tits in this way. He found that the birds would attack a pencil pushed through the bars of their cage and that they would flee from a bright light. When the light and the pencil were presented at the same time the birds tended to perform threat displays. Other threat displays could be elicited by presenting the pencil outside the cage so that direct attack was thwarted. Thus motivational conflict and thwarting of attack seem to generate threat displays.

HOW SIGNALS ARE MODIFIED DURING EVOLUTION

Although signals started off as incidental movements or responses such as those shown in motivational conflict, during their evolution they have become modified by selection to improve their effectiveness as signals. For example the courtship movement in which male ducks preen their wings has been emphasised in some species by the evolution of conspicuous bright coloured feathers on the wing, towards which the male points his bill during courtship preening. An extreme case is the mandarin duck (*Aix galericulata*) which has some of its wing feathers modified to form a bright orange 'sail' which is erected during courtship preening. The ancestral preening movement is reduced to a quick turn of the head so that the bill points to the orange sail.

The term *ritualisation* is used to refer to the evolutionary modification of movements and structures to improve their signal function. Thus the ancestral preening movement of ducks has become ritualised in species such as the mandarin duck. The changes that occur during ritualisation include the following: The movements tended to become highly stereotyped, repetitive and exaggerated; often the movements are emphasised, as in the mandarin duck, by the development of bright colours on the body. It is of course not possible to observe these evolutionary changes taking place and the evidence that they have occurred comes from comparative studies of displays in closely related species. In the example shown in Fig. 11.4 the probable ancestral movement is pecking at pieces of food on the ground during courtship (perhaps originally a displacement activity). This pattern is seen in present day jungle fowl (Fig. 11.4). In other species of galliform bird the display is ritualised; for example in pheasants and peacocks it is emphasised by the evolution of a large tail and the original ground pecking movement is reduced to a stereotyped bobbing of the head or pointing of the beak.

We have said that ritualisation occurs because it improves the signal function of a display. What exactly does this mean? As we saw in chapter 6 some displays used by males during courtship may be

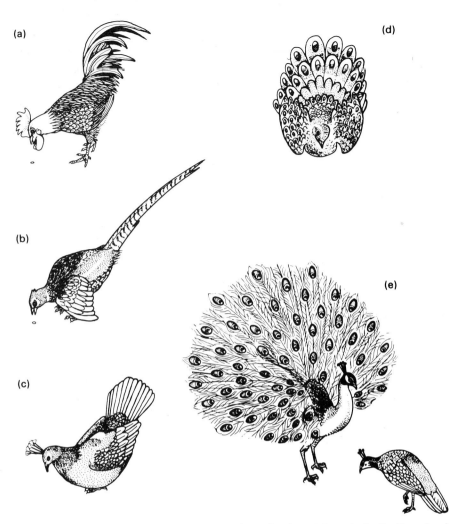

Fig. 11.4 Comparative evidence for the origin and ritualisation of a display. The ground pecking display of phasianid birds. (a) The least ritualised form is shown in the male domestic fowl (*Gallus*). It scratches the ground with its feet and pecks at food or small stones (perhaps originally a displacement activity). This attracts the female. (b) The male ring-necked pheasant (*Phasianus colchicus*) attracts females by means of a similar display. (c) The impeyan pheasant (*Lophophorus impejanus*) and (d) the peacock pheasant (*Polyplectron bicalcaratum*) both emphasise the pecking display with rhythmic bobbing of the tail or the head. (e) The peacock (*Pavo*) shows little of the ancestral movement. The male spreads his enormous tail and points his beak towards the ground (redrawn from Cullen 1972).

exaggerated and very elaborate as a result of sexual selection: females may exhibit a preference for elaborate displays. While this hypothesis may account for the ritualisation of some courtship signals, it cannot be a general explanation of ritualisation, since other displays such as those used in threat also appear to have undergone ritualisation. There are three hypotheses about the selective advantage of ritualisation. We will illustrate them mainly with reference to threat displays, although they could equally well be applied to other signals.

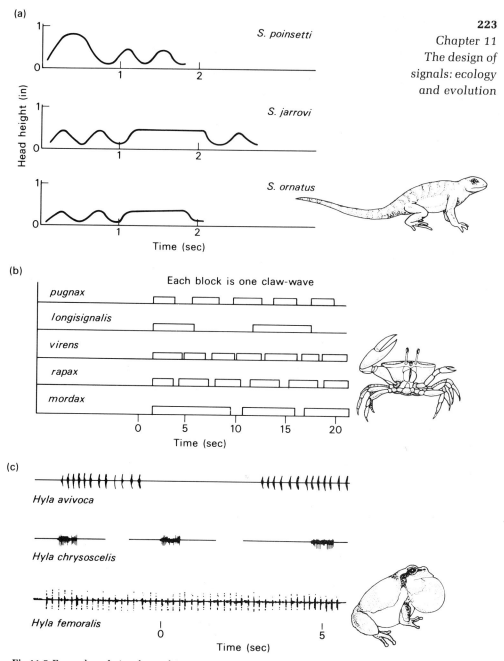

Fig.11.5 Examples of signals used in species recognition. (a) Lizards of the genus *Scleroporus* have species specific head bobbing patterns used in courtship and threat. The graph shows the head height as a function of time in the bobbing displays of three species. (b) Male fiddler crabs (*Uca*) attract females by waving their enlarged claw. Each species waves in a different way. The graph to the left shows the waving patterns of four species. A block represents a complete wave of the claw. (c) Tree frogs (*Hyla*) have species characteristic calls. Females are attracted only to males of their own species. The oscillogram traces to the left show the gross temporal characteristics of each of three species' calls. There are also species characteristic differences in the fine structure of calls that are important in species recognition.

224

(a) Reduction of ambiguity

According to this view ritualisation is the result of the selective advantage to signallers in reducing the chance of confusion between their various displays (Cullen 1966). Thus a ritualised threat signal, by its exaggerated stereotyped and repetitive nature clearly states 'I am about to attack' and not 'I am very frightened' or 'I am a sexy male'. In fact (as Darwin first pointed out) threat and appeasement signals are often extreme opposites: a threatening dog stands erect while a fearful dog crouches near the ground in appeasement. The principle of reduction of ambiguity can also be illustrated with reference to courtship signals. Often the displays of closely related species are clearly distinct so as to minimise the chance of confusion between species (Fig. 11.5).

Increasing the clarity of signals by increasing their stereotypy may at the same time reduce the amount of information they convey about the actor. The exact structure and movement pattern of an ancestral threat signal may have reflected the precise balance between aggression and fear in the actor, but a stereotyped ritualised signal probably conveys less information about the actor's state. Morris (1957) called this stereotypy 'typical intensity' and he viewed the loss of accuracy of information as the price paid by the actor for reduction of ambiguity. This assumes that it is advantageous for the actor to communicate information about its internal state, but in fact just the opposite may often be the case. As pointed out in chapter 5, if two animals are contesting a resource by means of ritualised signals, it will not pay either to reveal its exact balance between attack and fleeing until the last possible moment. In fact stereotypy of displays may have evolved precisely *because* it reduces the information available to reactors about the actors internal state. This leads us to the second hypothesis about ritualisation, which emphasises the use of signals by actors to control the behaviour of reactors.

(b) Manipulation

Selfish gene theory leads us to expect that social interactions will rarely be harmonious. Usually the actor and reactor will benefit to different extents from an act of communication and both will be selected to maximise their own benefit. This view suggests a new hypothesis of ritualisation. Consider, for example, the evolution of a threat signal. Actors alter the behaviour of reactors (cause them to retreat) by threat. Natural selection will favour actors whose signals are most effective in causing reactors to retreat and these signals are likely to be ones that are exaggerated, conspicuous and repetitive.

At the same time selection on reactors will favour resistance to exaggeration and bluff. Greater resistance will lead to even greater exaggeration in an evolutionary race between actor and reactor. The outcome of the race is what we observe as ritualisation. (Perhaps a useful analogy might be drawn between the process of ritualisation and the development of exaggerated and repetitive advertising claims to overcome the sales resistant public).

The term 'manipulation' may seem to be rather emotive, but it is a good description of the effect of ritualised signals on reactors in many kinds of communication. A striking example is the effect of a young cuckoo on its foster parents. By begging in the nest, the cuckoo manipulates the parental behaviour of its foster parents to the advantage of cuckoo genes and the manifest disadvantage of foster parent genes. A similar kind of manipulation in normal offspring parent communication has been proposed by Bob Trivers (1974). He pointed out that a young animal is more closely related to itself than it is to its siblings. Therefore it ought to be more interested in its own survival than in the survival of its brothers and sisters. The parents, however, are usually related equally to all their offspring and are therefore expected to allocate their resources equally between their children. (There are obvious exceptions to this rule, for example some birds of prey in which the brood hatches asynchronously and the parent only feeds the smallest young if food is plentiful.) Trivers therefore predicted that there would be conflicts of interest over allocation of parental resources. Each young would try to acquire more than its 'fair share' of resources while the parents would try to allocate resources equally. (This is analogous to the parent offspring conflict described in chapter 10.) Therefore begging by young for food and warmth may contain an element of exaggeration in an attempt to persuade parents to give extra food. Some indirect support for this idea comes from a study of begging by hand reared nestling great tits. If the birds were fed (by hand) on demand they continued to remain dependent on the foster parents for longer after fledging than if their begging was not rewarded at once by food. The implication is that the young 'attempt' to acquire food from their parents for a long period after fledging, even if they are potentially capable of feeding themselves (Davies 1978b).

(c) Honesty

In chapter 5 we discussed the idea that threat signals are often accurate indicators of the size or strength of the actor. We suggested that only reliable cues such as pitch of voice would be used by reactors to assess their rivals, so that unreliable signals will have disappeared during the course of evolution. Amotz Zahavi (1979) has suggested that ritualisation is the result of selection by reactors for reliable signals. During evolution, he argues, reactors are selected not just to become sales resistant but to discriminate between signals on the basis of their reliability as indicators of an individual's strength, size, motivation to fight, parental ability and so on. One way to select for reliability is to select for stereotypy and repetition of the display. In the same way that human judges assess the difference between athletes by asking them all to perform exactly the same task under the same conditions, reactors will be better able to judge differences between actors by selecting for stereotyped displays. The crucial feature of ritualisation according to Zahavi is not just that displays become more stereotyped and repetitive, but that the stereotypy provides a uniform background against which subtle differences between individuals are emphasised. Zahavi might argue that when the male zebra finch hops towards a potential partner in his stereoptyped courtship dance, the female is assessing the male's agility, as an indicator of his ability to gather food for the young.

Which of these views is correct? It is difficult to design experiments to distinguish between them since they are ideas about evolution over an historical time scale, but they could all be correct for different stages in ritualisation. Signals may initially become ritualised because of the benefit of reduced ambiguity, then selection on actors to manipulate reactors may have led to greater exaggeration and stereotypy, and finally sales resistance and discrimination by reactors may have led to displays, particularly those used in assessment, being performed in a way which allows assessment of differences between actors. Signals which do not allow reliable assessment may gradually lose their effectiveness, fall into disuse and be replaced by new ones.

The three hypotheses can to some extent be assessed by indirect tests. As mentioned in chapter 5, Caryl (1979) analysed the threat displays of several species of birds and found that in general threat signals are rather poor predictors of ensuing attacks. This would be more consistent with the 'manipulation' hypothesis, which predicts that actors should give away little information about their motivational state, than with the 'reduction of ambiguity' hypothesis, which predicts that threat signals should communicate with maximum clarity the future behaviour of the actor. However Hinde (1981) suggests that this analysis may be too simple. He points out that some of the displays considered by Caryl are good predictors of 'attack or stay put' or 'flee or stay put'. This might indicate that the actor signals a conditional strategy (chapter 5) and that its future behaviour depends on the reactor's response. This interpretation requires an advantage for signallers to communicate intention to retreat (otherwise it would pay to signal high probability of attack all the time in the hope of deterring the rival). One suggested advantage is that high signalling probability of attack is more likely to evoke vicious retaliation, so it is only worth taking this risk when the benefit of staying is high.

VERY COMPLEX SIGNALS: BIRD SONGS

Perhaps the most puzzling signals from the point of view of design are the very complex song repertoires of passerine birds. A male nightingale or mockingbird may produce hundreds of different phrases in a seemingly endlessly varying outburst of song. If signals convey messages such as 'keep out' 'I am a male of species A' and so on, why the need for such extraordinary elaboration? As E. O. Wilson (1975) neatly put it, the signal seems redundant to the point of inanity. Although there are still many unanswered questions, the evidence suggests that song complexity is important in both mate attraction and territory defence.

(a) Attracting males

As we saw in chapter 6, the very complex song of the sedge warbler is a product of sexual selection: females are more attracted to males with the largest repertoires. A similar interpretation of complex songs is suggested by Don Kroodsma's (1976) observation that female canaries are more readily stimulated to build nests when they hear a complex song repertoire than when they are given an artificially simplified repertoire. Older male canaries have larger repertoires, so females may use song complexity as a way of assessing the age and experience of potential males.

Competition between males for mates is more intense in polygamous than monogamous animals (chapter 7), so it is to be expected that sexually selected displays will be more highly developed in the former. Kroodsma (1977) found this pattern in the songs of North American wrens: monogamous species such as the rock wren (*Saltinctes obsoletus*) have relatively simple songs while the polygamous marsh and winter wrens (*Telmatodytes palustris*, *Troglodytes troglodytes*, respectively) have very complex songs. Song elaboration in wrens is therefore probably a result of sexual selection.

(b) Territory defence

Many species use song not only for attracting mates but also for proclaiming territory ownership, and here complex repertoires may also play a role. This was shown in an experiment in which great tits were removed from their territories and the empty spaces were occupied with loudspeakers broadcasting song. Territories occupied with song repertoires were invaded more slowly than those occupied with loudspeakers playing a single song type (control territories with no song were invaded fastest of all) (Krebs *et al.* 1978). It is not yet known why repertoires are better at keeping out intruders, but one suggestion is that they are a form of deception. Intruders are looking for an empty territory, or at least a sparsely populated piece of habitat. Residents deter intruders by singing, and a variety of songs may be a more effective deterrent because it creates the impression of a habitat crowded with many singing males. Great tits tend to change from one song to another at the same time as changing perches, which enhances the illusion of many birds. Selection will act on intruders to spot the deception and the result will be an evolutionary race as we discussed with reference to parents and offspring. The outcome of the race may sometimes be very elaborate song repertoires.

In summary, the influence of reactors on the design of signals leads to ritualisation—an increase during evolution of exaggeration, repetition and stereotypy. Although ritualisation may sometimes occur because of the benefits of reducing ambiguity in signals, it may often be the product of an evolutionary chase in which actors are selected to increase the effectiveness of their signals and reactors are selected to become sales resistant.

Summary

Communication in animals occurs when one individual uses specially designed signals or displays to modify the behaviour of others. The design of signals is influenced both by ecological constraints and by their effectiveness in modifying the behaviour of reactors. The habitat can influence the effectiveness of different sensory channels of communication (e.g. scent versus visual signals) and the exact form of signals within a sensory channel. The latter point is illustrated by differences between the songs of birds living in different kinds of vegetation. As signals evolve selection improves their effectiveness by making them stereoptyped, repetitive and exaggerated. There are three hypotheses to explain why these changes make signals more effective; (a) reduction of ambiguity, (b) manipulation and (c) honesty.

Further reading

Catchpole (1979) is a good general review of bird song including discussions of song repertoires and song mimicry.

Cullen (1972) is a good summary of the older ethological literature on animal signals.

Dawkins and Krebs (1978) develop the idea that communication is a matter of manipulation by actors of reactors. They contrast this view with the 'classical ethological view' that signals evolve for efficient transfer of information.

Hinde (1981) criticises Dawkins and Krebs and questions whether their view is very different from that of earlier ethologists.

Andersson (1980) discusses the idea of threat displays falling into disuse because they are mimicked by bluff.

Wiley & Richards (1978) provide a technical review of ecological constraints on sound signals and Bowman (1981) describes how the songs of Darwin's finches are adapted to their habitats.

The book by Hailman (1977) is an excellent review of visual displays in animals.

Chapter 12. Co-Evolution and Arms Races

In the last three chapters we have seen that what at first appear to be examples of harmonious co-operation within a species usually involve conflicts of interest between selfish individuals. This is true even for the most extreme examples of co-operation such as reproductive behaviour in the social insects. We will now broaden our outlook a little to consider some cases of co-operation between members of different species. These provide some of the most extraordinary observations in the whole of animal behaviour.

Small birds hop into the open jaws of crocodiles and pick between their teeth; the birds find food and the crocodiles get their teeth cleaned! Some fish allow other species of fish to eat parasites from their skin; they gain through having the parasites removed and the cleaners get a tasty meal. Acacia plants provide nest sites for ants and also food secreted from nectaries at the base of their leaves. In return, the ants protect the plants against attack by herbivorous insects.

The term 'co-evolution' is often used to describe interactions like these where adaptations in one species have evolved in relation to adaptations in another. The examples above all involve 'mutualism', inter-specific relationships where each partner gains a net benefit. However just as conflicts of interest underlie acts of apparent co-operation within a species, so we should expect conflict to lie at the heart of interspecific relationships. Even if both species gain a net benefit from the association a closer look may reveal an evolutionary arms race, like that between a predator and its prey, with each partner selected to increase its benefit even more, at the expense of the other.

We will restrict our discussion in this chapter to one example of co-evolution, namely the relationships between plants and animals, where the flowers are visited by bees which bring about pollination and then later, when the seeds have set, these are dispersed by birds. We have chosen this example not only because it illustrates so clearly the conflicts of interest between plant and animal even in a mutualistic relationship, but also because it brings together some of the ideas on optimality that we discussed earlier in the book. At the end of the chapter, when the flowers have been pollinated and the seeds dispersed, we will have a more general look at the evolution of arms races.

Flowers and bumblebees

It is said that the British Empire once owed its power and wealth to bumblebees. Britain's power resided in her navy, the navy subsisted on beef, beef came from cattle, cattle ate clover and the clover was pollinated by bumblebees. From the plant's point of view the bee is a flying penis, carrying pollen from one plant to the ovaries of another. The reward for the bees' trouble is a drink of nectar. Selfish gene theory leads us to predict that this relationship will not, however, be one of harmonious mutualism. Instead we predict that the bees will be selected to exploit the nectar optimally even if this is at the expense of efficient pollen transfer. On the other hand, the plants will evolve to exploit the bees; it costs valuable resources to produce the nectar and plants will be selected to pay the minimum price for efficient pollination. Let us look at this sytem in detail and see whether these predictions are borne out and, if so, how the conflict is resolved.

THE PLANTS: SELECTION FOR OUTCROSSING

Mary Price and Nick Waser (1979) suggested that for sexually reproducing plants the healthiest offspring will result from matings between individuals that are not too similar in genetic make up, but not too dissimilar (Fig. 12.1a). If the mating individuals are similar genetically then lethal homozygotes will occur in the offspring resulting in inbreeding depression. On the other hand, if the mating individuals are very dissimilar then there would be outbreeding depression because of the disruption of favourable gene combinations, adapted to local conditions. This same model had been put forward earlier by Pat Bateson (1978) for mate selection in animals (see chapter 13).

Price and Waser tested this idea by hand-pollinating flowers with pollen brought from different distances. Plants living close together are likely to be closely related while those far apart are more likely to be genetically dissimilar. They scored success in terms of the number of seeds set per flower and found that intermediate distances of pollen transfer resulted in greater success than short or very long distances (Fig. 12.1b). The curve was rather flat topped and the optimal outcrossing distance was between 1 and 100 m.

They then measured the movement of pigmented pollen, carried by bees, to determine the actual outcrossing distance achieved in the wild. The mean distance was about 1 m, though some pollen was transferred as much as 10 m. These observed values are towards the lower end of the optimum predicted by the experiments. It is not obvious why this is so, but we should recognise that pollen transfer is very difficult to measure in the wild and that the plant could still discriminate between pollen after it has been transferred (Lewis 1979).

The two important conclusions from Price and Waser's work are that the plants will benefit from outcrossing and that outcrossing in nature is achieved by the movements of pollinators such as bees.

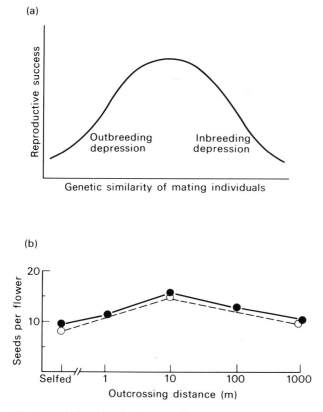

Fig. 12.1 (a) The idea is that reproductive success will be greatest if the mating individuals are not too similar genetically, but not too dissimilar. (b) Effect of outcrossing distance on seed set per flower in hand-pollinated *Delphinium nelsoni*. The two lines refer to different experiments (from Price & Waser 1979).

THE PLANTS AS MANIPULATORS AND THE BEES AS
OPTIMAL FORAGERS

How do the plants get the bees to perform efficient outcrossing?

(a) Flowering times

Pollen transfer will be most efficient if the bee flies from one plant to another of the same species. If, for example, a bee flew from a rhododendron to a willow flower then pollen would be wasted. One way in which plants might encourage flower constancy by the bees is by synchronous flowering within a species and by sequential flowering during the year of different plant species. In effect different plant species are competing with each other for the attention of the pollinators and so it pays a species to flower when others are not (Heinrich 1979). A species may have lower reproductive success at times when its flowering overlaps with other species (Waser 1979).

(b) Individual specialisation

Despite this system of sequential flowering, in any one place there are still usually several species of plants in bloom at the same time. A visit to a field on a sunny day may show that one species of bumblebee is visiting many species of flowers. However, a closer look would reveal that individuals were specialising.

When Bernd Heinrich (1979) marked individual bumblebees (*Bombus* spp.) with paint spots and followed their foraging trips, he found that they each had a favourite species of flower which he called their 'major'. At any one time one individual might be majoring on willowherb, another on clover, another on carrot, and so on. Heinrich found that such individual specialisations could last for a month or more. This is obviously good for the plant because it promotes efficient pollination, but why is it good for the bee?

The answer is that different species of flowers have different morphology and consequently require different techniques for reaching the nectar. In some the bee has to enter a funnel, in others it has to push its way under a hood formed by the sepals (Fig. 12.2). Heinrich followed newly emerged bumblebees on their first foraging trips and found that their feeding efficiency improved with experience on a particular flower species because they had to learn the technique for discovering the nectar (Fig. 12.3). The plants may have evolved these complicated entrances and unique positions of the nectar to force the bees to specialise. Generalisation is prevented because the bee would have to learn the skills involved for each species of flower and would have to be continually changing its foraging tactics. Different flower shapes and colours may have evolved as aids to help individual bees remain species constant.

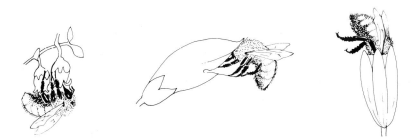

Fig. 12.2 Different species of flowers have different morphologies and so the bee needs different techniques to get at the nectar.

As well as specialising at a major flower, each bee also visited one or two other species or 'minors'. Minoring may enable the bees to keep track of any changes in flower profitability. This idea was tested by Heinrich. When he removed the major flowers or enriched the nectar of a minor species by adding sugar, the bees quickly switched to specialise on the minor.

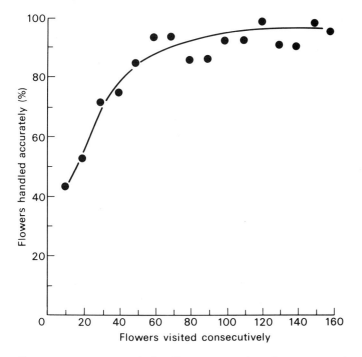

Fig. 12.3 Improvement in the handling accuracy of *Bombus vagans* workers foraging at jewel-weed flowers *Impatiens biflora*, starting with the first flower encountered in the bees' foraging career (from Heinrich 1979).

How does an individual bumblebee come to choose its major flower? Unlike the mass communication systems in honeybees (*Apis mellifera*), where successful individuals inform others about the identity and location of good food supplies (von Frisch 1967), each individual bumblebee (*Bombus* spp) has to learn for itself which flowers are the most profitable sources of nectar. Within a field some species of plants will produce more nectar than others. The first bees to arrive in the field will forage at these flowers but the more individuals that major on them, the more the nectar will be depleted. Eventually the level of nectar will be depressed so much that it will now pay the next bees to arrive to choose another species of flower.

It is clear, therefore, that the best major flower for any one individual must depend on what all the other individuals are doing. If there is no territoriality, and ideal free conditions prevail (see chapter 5), then the result should be that all species of flowers are depleted down to the same level of profitability. This is what Heinrich observed (Fig. 12.4). If the nectar rewards of one species increased above that of the others then it would pay individuals to forage on it until its profitability dropped to the same level as the other species.

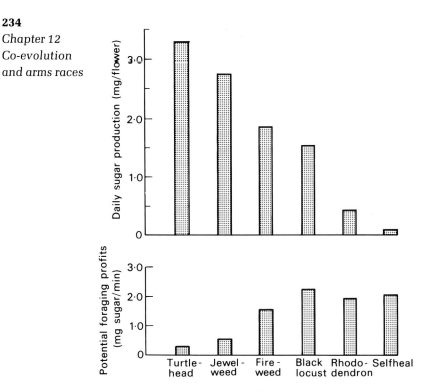

Fig. 12.4 Daily sugar production of individual flowers (top graph) varies in different species. Because of the depletion by bees of flowers with high nectar production, many of the flowers end up offering roughly similar food rewards per given foraging time (bottom graph). (From Heinrich 1979.)

(c) Number of visits

A plant will get more pollen transferred if a bee visits it several times rather than just once. Whitham (1977) discovered that the desert willow (*Chilopsis*) secretes its nectar in such a way that bumblebees are encouraged to visit it more than once in a day.

A cross section of the nectar tube reveals a pool of nectar in the centre which is easy for the bee to reach and then, radiating out from the pool are several narrow grooves. The nectar in these grooves is much harder for the bees to extract. The result is that when a bee starts drinking at a flower it gets rewards quickly at first and then more slowly as the pool nectar is used up (Fig. 12.5). The bee's problem is exactly the same as that of the copulating male dung fly discussed in chapter 3, namely when to give up on a curve of diminishing returns. The best point to give up will depend on the expected rewards from leaving and flying to another flower.

Whitham found that in the early morning all the flowers had a lot of nectar and the optimal strategy for the bees to maximise nectar intake was to go rapidly from flower to flower just drinking the pool nectar. By feeding in this way they got 12.3 cal per min, whereas if they had

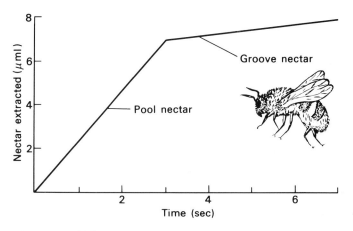

Fig. 12.5 Bumblebees extract the pool nectar of desert willow flowers at a faster rate than the groove nectar (after Whitham 1977).

stayed at each flower until the groove nectar was depleted as well, they would only have got 9.9 cal per min. Later in the day the nectar levels were lower and the bees went round the willow flowers again. This time they stayed longer at each flower and drank the groove nectar as well, as expected from optimal foraging theory. We adopt the same foraging strategy ourselves when hunting for blackberries. At the start of the season, when berries are abundant, we move quickly from bush to bush picking the largest and most juicy fruits because the expected gain from moving is high. Later on, when berries are scarce and the expected gain from moving is low it pays to stay longer at each bush in search of the last few (see Box 3.2, chapter 3).

(d) Movements on a plant

There is a further feature of plant design that appears to be an adaptation for efficient pollen transfer.

In plants like *Epilobium*, and *Delphinium*, the flowers are arranged in a spiral round a vertical spike. They are protandrous hermaphrodites; the young flowers at the top of the spike are males which shed pollen but have unreceptive stigmas while the older flowers at the bottom are females, with receptive stigmas and no pollen. The plant's goal is to get pollen transferred from its top flowers to the female flowers of another plant and to receive pollen from another plant onto its female flowers at the bottom of the spike. From the plant's point of view, therefore, it would be best if the bee started at the bottom and moved upwards. In ths way pollen would be transferred from the top of one plant to the stigmas at the bottom of the next one (Fig. 12.6).

Graham Pyke (1978, 1979b) found that the plants manipulate the bees into moving in this way by secreting most nectar in the flowers at the bottom of the spike. The best foraging strategy for the bee is,

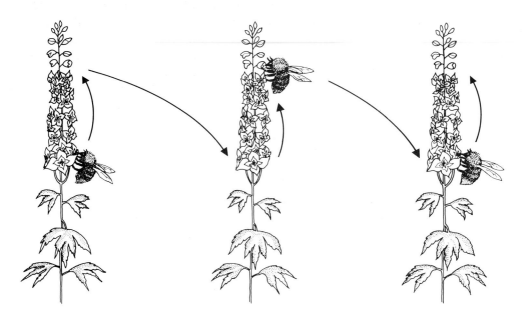

Fig. 12.6 When foraging on *Delphinium* the bees start at the bottom flowers, which contain the most nectar, and move vertically upwards. This is good for the plant because it promotes outcrossing.

therefore, to start at the bottom flowers where the nectar is richest and move upwards. By always moving upwards the bee avoids revisiting flowers that it has just depleted. It will experience a curve of diminishing returns as it goes up the spike and encounters flowers with less and less nectar. Eventually it will reach a point where the expected gain from staying drops to that expected from moving on. At this point it will pay the bee to leave and fly down to the bottom of the next plant.

We are still left with the problem of why the flowers are arranged in a spiral pattern around the spike. Pyke suggests that a bee never carries sufficient pollen to fertilise all the female flowers on a plant. Therefore it is not in the plant's interest for the bee to visit all the female flowers and drink all the nectar because only the first few will get the benefit of pollination. It is possible that the spiral arrangement forces the bee to miss some flowers because moving sideways round the spike may be more costly than moving upwards to the nearest flower. Observations show that bees tend to move vertically upwards, missing out some flowers on the spiral, rather than moving sideways so as to visit every flower on the spike.

CHEATERS IN THE SYSTEM

Given that some plants produce nectar to attract bees it is not surprising to find that others are cheaters, mimicking the nectar producers but not secreting any nectar. They thus avoid the cost of producing nectar and get their pollen transferred free.

In the bogs of Maine, U.S.A., the grass pink orchid (*Calopogon pulchellus*) is a cheater. It mimics the rose pogonia (*Pogonia ophioglossoides*) which is a nectar producer. Of course selection will act on the bees to spot this deception because their visits to the orchids will be time spent foraging for no rewards. As a counter-measure the orchid makes it difficult for the bee to learn that it is a cheater by being very variable in colour.

The orchid has the problem of having its pollen transferred by the bee to the species it mimics rather than to another orchid. It has got round this by arranging the stigmas and anthers in a different way from the rose pogonia. In *Pogonia* the pollen is brushed onto the bee's head while in *Calopogon* it goes onto the bee's abdomen. So even if a bee does fly from the orchid to the pogonia, the orchid's pollen will still remain in place until the bee visits another orchid (Heinrich 1979).

Not only do plants sometimes cheat the bees, but bees sometimes cheat the plants. Some *Bombus* spp. bite through the petals at the base of flowers instead of entering through the funnel of the corolla. This is advantageous for the bee because it can get the nectar more rapidly but it is against the interests of the plant because no pollen is picked up.

Birds, fruit and seed dispersal

Once pollination has been achieved and the flowers have set seed, the plant's next problem is to get its seeds dispersed. Many species rely on birds for dispersal and one trick they use to attract them is to cover the seeds with fruit. The bird eats the fruit and then, sometime later, the seed passes through its digestive system unharmed.

Fruit is selected to be attractive and so we have the strange situation of a prey that 'wants' to be eaten by a predator. David Snow (1971) has contrasted this with the relationship between bird predators and insect prey which is selected to be difficult to find and capture. Table 12.1 summarises some of the interesting consequences of these different predator–prey systems.

Birds are not the only animals that are attracted to fruit. It is also attacked by microbes which cause the fruit to rot. Dan Janzen (1979) has put forward an ingenious explanation for why fruit rots. He suggests that this is the way microbes prevent larger organisms from eating their food supply. By manufacturing ethanol they render the fruit distasteful to vertebrate competitors. Birds will be selected to avoid rotting fruit because if they eat it they get drunk and become susceptible to predators. Similarly, Janzen suggests that bacteria putrefy carcasses and fill the meat with noxious compounds, like amines, to prevent vertebrates eating them. Incidentally, some vertebrates have evolved resistance to this poisoning. They eat the rotten fruit and then retain the microbes in their guts to digest the food for them! Perhaps another counteradaptation by vertebrates is to possess an enzyme capable of breaking down alcohol (alcohol dehydrogenase). In man this is a useful preadaptation for recovering from late night parties but it may have originated because our ancestors ate fruit.

Table 12.1 Contrasting relationships between predators and prey, comparing birds which eat insects with those that eat fruit (from Snow 1971).

Insects	Fruit	
Selected to be difficult to find and capture	Selected to be easy to find and capture	
↓	↙	↘
Diverse escape mechanisms	Fewer niches	Different species of fruit compete for birds' attention
↓	↓	
Many 'niches' for predators	Fewer species of frugivorous birds	
↓		↓
Many species of insectivorous birds		Staggering of fruiting times
		↓
Bird needs large hunting area, undisturbed by others, and both parents needed to feed young.		Ensures birds have a continuous food supply
Therefore; territoriality, monogamy.	Social feeding, no territoriality	
	Female able to feed young by herself	
	Male emancipated from parental duties	
	Therefore; lek displays, promiscuity, bright plumage in males	

Fruit will also be attacked by animals which destroy the seeds rather than disperse them. Small mammals, like mice, may crush the seeds with their teeth and so plants protect their fruit with spines on the stems or by displaying them so that only birds can enjoy easy access.

A particularly beautiful example of how plants have evolved adaptations against seed predators is a study by Ron Pulliam and Marina Brand (1975) in the grasslands of south-eastern Arizona. They noticed that seeds produced in the summer, after the winter rains, are round and smooth in shape. The main seed predators at this time of year are ants which find smooth round seeds difficult to pick up with their mandibles. Finches find smooth seeds easy to handle, because they can turn them round and round inside their mandibles for husking, but in the summer the finches are mainly eating insects. Seeds produced in the winter, after the summer rains, are very different in morphology. They are rough and have many projections. Ants would find these seeds easy to carry but in winter the ants are not active. The main seed predators are now the finches and they find rough and hairy seeds rather difficult to handle because they get stuck in the grooves inside the mandibles and are difficult to husk. Therefore the morphology of the seeds is nicely adapted to reduce the risks of predation at each time of the year (Fig. 12.7).

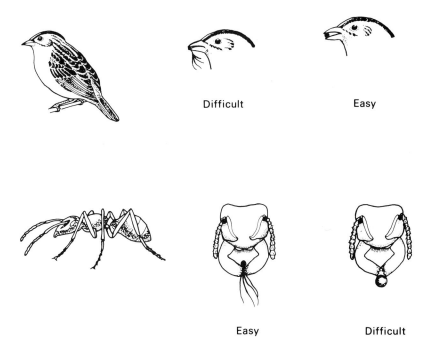

Difficult   Easy

Easy   Difficult

Fig. 12.7 Finches find hairy seeds difficult to handle but round seeds are easy. In winter, when the finches are the main seed predators, most seeds produced are hairy. Ants find hairy seeds easy to carry but have difficulty with smooth round seeds. In summer, when the ants are the main seed predators, most seeds produced are round (after Pulliam & Brand 1975).

Some plants rely on seed predators for dispersal. Although they will have some of their seeds destroyed many will escape, especially when the animals hide them away in stores. The seeds of the pinon pine are stored by nutcrackers and jays (Vander Wall & Balda 1977; Ligon 1978). The birds carry the seeds to sheltered south facing slopes which do not get covered in deep snow and so are both good sites for future retrieval and good for germination. Many seeds are not recovered by the birds, especially during mild winters when they do not need to use all their stores. In some mild winters the nutcrackers only eat a third or a half of their seed supplies and so many escape predation and germinate.

Plants that rely on seed predators for dispersal pay the cost of having some of their seeds destroyed by the predator while plants that rely on fruit as a lure pay the cost of making the fruit.

Arms races

Examples of co-evolution like those we have discussed above may result in a net benefit to each partner but it is clear that there are conflicts of interest. The plants have evolved tricks to manipulate the bees to transfer their pollen and the birds to disperse their seeds, and they will be selected to pay the minimum price in nectar and fruit for the maximum benefits. On the other hand, the bees and birds will be selected to exploit their food optimally, to obtain the maximum amount of nectar and fruit per unit foraging effort. Conflicts of interest are particularly apparent when the plants cheat the animals and vice versa.

Perhaps these systems are better viewed as evolutionary escalations of counter-adapatations or 'arms races' (Dawkins & Krebs 1979) like those between predators and their prey.

PREDATORS AND PREY

The relationship between bees and flowers and birds and fruit are mutualistic because during the arms race each partner gains a net benefit. In arms races between predators and prey, however, only the predator gets a benefit. During evolution we would expect natural selection to increase the efficiency with which predators detect and capture their prey. On the other hand we would also expect selection to improve the prey's ability to escape. What will be the result of such an arms race? The complex adaptations and counteradaptations we see between predators and prey are testament to their long co-existence and co-evolution. But why don't either predators become so efficient that the prey are driven to extinction or the prey evolve to be so good at escaping that the predators become extinct?

Models derived to tackle this question have usually considered just two species, namely one predator and its prey. Although this is obviously too simple to reflect exactly the complex food webs we see in nature, it at least helps an understanding of the kinds of problems involved (Slatkin & Maynard Smith 1979). A useful model to help us think about predator–prey co-evolution is the following derived by Rosenzweig & MacArthur (1963). If we let $P(t)$ represent the number of predators and $V(t)$ the number of victims (prey) at time t, then we can write the following equations for population changes.

For the victims, $\dfrac{dV}{dt} = f(V) - Pg(V)$

where $f(V)$ describes the growth of the prey population in the absence of predators and $g(V)$ describes the number of prey eaten by an individual predator in relation to the number of prey available (the predator's 'functional response').

For the predators, $\frac{dP}{dt} = -dP + cPg(V)$

where c is the conversion rate of prey to predators and d is the death rate of the predator.

Rosenzweig and MacArthur presented this predator–prey interaction graphically as follows. If there are no predators then the prey population will increase to some maximum density, K, set for example by their food limit. There must also be some lower limit to prey density below which the population will become extinct because individuals rarely meet to reproduce and chance fluctuations in the environment reduce their numbers to zero. Between these two limits, for each density of prey, there must be a maximum predator density that can be supported without either a decrease or an increase in the prey population. We can therefore plot a prey curve, or isocline, at which

$$\frac{dV}{dt} = 0$$

(see Fig. 12.8a). Below the curve the prey population will increase and above it will decrease. By a similar argument we can plot a predator isocline at which

$$\frac{dP}{dt} = 0$$

(Fig. 12.8b). Below some threshold prey density, X, the predator population will decrease because it cannot find enough food, while above the threshold it will increase up to some carrying capacity. The reason for the threshold is that the predator equation must give positive, negative or zero population growth.

If we now plot these two curves on the same graph, three conditions may occur (Fig. 12.9). In all cases an equilibrium will come about where the two curves intersect, where

$$\frac{dV}{dt} \text{ and } \frac{dP}{dt}$$

are both zero, i.e. prey and predator populations constant. Away from this equilibrium point the populations will move in the four ways indicated by the four quadrants A, B, C and D. In quadrant A both predator and prey are increasing; in B the predator increases and the prey decreases; in C both species decrease and in D the prey increases while the predator decreases. The arrows in Fig. 12.9 are vectors which indicate population changes in these four quadrants and the graphs on the right hand side of the figure indicate population changes with time. The three cases represent three levels of predator efficiency.

1 If predators are relatively inefficient the system will move towards the stable equilibrium.

2 If predators are very efficient there will be extreme fluctuations in

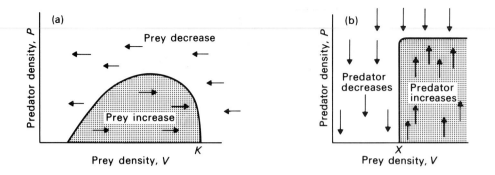

Fig.12.8 (a) The prey isocline,

$$\frac{dV}{dt} = 0;$$

in the shaded area below the curve the prey population increases and in the white area above the curve it decreases. It is assumed that intermediate densities of prey, where turnover rate is greatest, can support the highest density of predators without decreasing.
(b) The predator isocline,

$$\frac{dP}{dt} = 0;$$

the predator population increases beyond a threshold prey density, X. For simplicity it is assumed that the predator isocline rises vertically up to the predator population's carrying capacity (from Pianka 1974; after Rosenzweig & MacArthur 1963).

numbers and eventually either the predator, the prey, or both, will go extinct.

3   If predators are moderately efficient there will be stable oscillations in population densities.

One way of using these graphs to help us understand arms races is to imagine how the ecological parameters could change during evolution. Natural selection on the predators will tend to make them more efficient at exploiting the prey and so move the predator isocline to the left. Natural selection on the prey will make them less vulnerable to predation and so tend to move the predator isocline to the right. From Fig. 12.9 it can be seen that if the predator becomes too efficient the system will be unstable. But since we observe many stable predator–prey systems in nature we have to explain how this could come about. Three hypotheses can be suggested to explain why predator–prey systems tend to be stable (Slatkin & Maynard Smith 1979).

(a) Prudent predation

Man has the capacity to be a prudent predator and avoid over-exploiting his food supplies to extinction. It has been suggested that

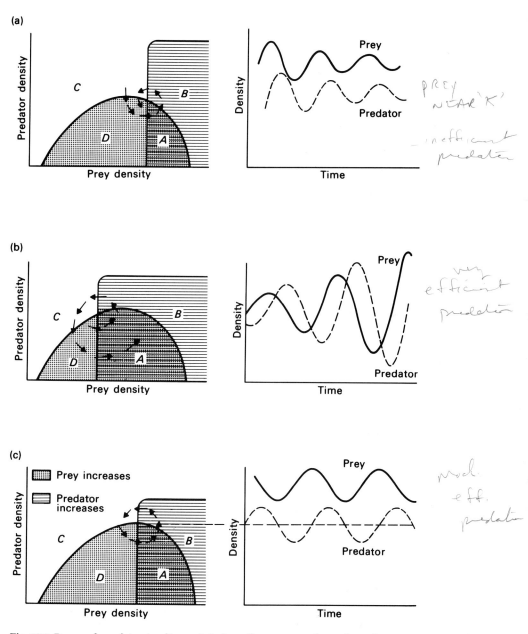

Fig. 12.9 Prey and predator isoclines plotted on the same graph to show three stability relationships. (a) An inefficient predator, that cannot exploit its prey until the prey population is near its carrying capacity. (b) A very efficient predator which is able to exploit even sparse prey populations. (c) A moderately efficient predator which begins to exploit its prey at some intermediate density. See text for further explanation (from Pianka 1974; after Rosenzweig & MacArthur 1963).

animal predators may likewise be prudent and avoid exploiting their prey too effectively so as to keep the predator isocline to the right of the hump in the prey curve in Fig. 12.9. The problem with this hypothesis is that it relies on group selection. In a population of prudent predators any individual that cheated and ate more than its 'fair share' would put more genes into future generations than prudent individuals (see chapter 1). Prudent predation could, however, evolve where an individual had exclusive use of a resource (e.g. by defending a territory) and so was saving food for its own future use rather than for the good of the population.

(b) Group extinctions

If group extinctions *were* common then the reason we see stable predator–prey systems in nature may be because all the unstable systems (e.g. those like Fig. 12.9b) have indeed gone extinct.

(c) Prey are ahead in the arms race

Predator–prey systems may be stable because the prey are always one step ahead in the arms race (i.e. the predator isocline will tend to be as in Fig. 12.9 (a) or (c) rather than (b). One hypothesis for why this might be so can be described as the 'life-dinner principle'; rabbits run faster than foxes because the rabbit is running for its life while the fox is only running for its dinner. The cost of a mistake is clearly greater for the rabbit. As Dawkins & Krebs (1979) put it, 'A fox may reproduce after losing a race against a rabbit. No rabbit has ever reproduced after losing a race against a fox. Foxes who often fail to catch prey eventually starve to death, but they may get some reproduction in first'. Therefore selection pressure will have been stronger on improving the ability of rabbits to escape than the ability of foxes to make successful captures.

In many cases the prey may also be ahead in the arms race because they have a shorter generation time than the predator and can therefore evolve at a faster rate. This will apply to examples such as weasels versus mice and birds versus insects but in some cases it is the predator which has the capacity to reproduce at a faster rate (aphids versus roses).

We may also ask why prey do not become so efficient at escaping as to drive their predators extinct. One hypothesis is that as predators become rare, because of increased prey efficiency, they exert little selection on the prey for further improvement. A second possibility is that the more efficient prey become, the more unlikely it is that new mutations will arise that allow further improvement.

Finally we should reinforce what we said at the start of this section, namely that although simple two species models can help us to think about predators and prey they are inevitably too simple. If a prey species became rare, due to excessive predation for example, the predator is likely to seek other prey species. Because adaptations for feeding on different prey are usually different, a predator is unlikely to be sufficiently specialised on any one species to drive it to extinction.

As we have seen so often in the book, arms races can also occur intraspecifically, such as the conflicts between workers and queens in the social insects (chapter 10) and other conflicts between parents and offspring (chapter 11). The gaudy colours and elaborate weaponry of many male animals is the result of an arms race within the species for the ability to fight for and attract females (chapter 6). At the heart of the conflict between male and female over sexual reproduction lies the primaeval arms race between microgametes and macrogametes. Microgametes were selected to seek out macrogametes for mating to parasitise their greater nutrient supplies. The macrogametes were selected to resist because they would have done better by joining other macrogametes. But the microgametes won this arms race because of the life–dinner principle. The cost of making a mistake for them was greater. If they mated with another microgamete the zygote would not have had sufficient reserves to survive. Therefore selection for the ability of small gametes to mate with large gametes would have been stronger than selection for resistance by the large gametes (Parker, Baker & Smith 1972, see chapter 6).

The most striking fact about intraspecific arms races is that they occur within the same gene pool (Dawkins & Krebs 1979). There is a conflict between parents and offspring (see chapter 10) but individuals who are now offspring will one day become parents. Males and females are usually in conflict but unless the genes for adaptations concerned with sexual conflict are on the sex chromosomes, they will have spent half their history in male bodies and half in female bodies. Worker and queen ants are in conflict but every female ant is genetically equipped to become either a worker or a queen. Which role it plays simply depends on its nutrition as a larva. If it becomes a queen it will be selected to invest in reproductives in the ratio one male to one female. If it becomes a worker it will be selected to invest in one male to three females (see chapter 10). It is not yet clear whether arms races such as these, which occur within the same genetic lineage, will have different consequences from those involving separate lineages, such as predators and their prey.

Summary

In some interspecific relationships each partner gains a net benefit. However underlying this so-called 'mutualism' there are conflicts of interest. One example is the co-evolution of flowers and their pollinators, the bumblebees. The plants manipulate the bees into performing efficient outcrossing by staggering of flowering times in different species, by having unique morphologies so as to force the bees to specialise, by the way they produce nectar and, for flowers arranged on spikes, by putting most nectar into the lower flowers on the stem. The bees maximise the nectar they obtain for the minimum costs. Sometimes the plants cheat the bees and vice versa. Plants may get their seeds dispersed by covering them in fruit or by relying on seed predators. Most examples of co-evolution are better considered as arms races like those between predators and prey. Predator–prey systems may tend to be stable because prey are usually ahead in the arms race.

Further reading

The book by Heinrich (1979) is a discussion of co-evolution between plants and bumblebees and is a fine example of how theory and experiments can help an understanding of behaviour in the field. Another good example of an arms race is brood parasitism; see the reviews by Payne (1977) and Yom-Tov (1980). N. G. Smith's (1968) study is a particularly fascinating account. The book edited by Gilbert & Raven (1975) contains some good studies of co-evolution; see especially the chapter by Gilbert on butterflies and plants. The books by Wickler (1968) and Edmunds (1974) are summaries of predator–prey adaptations and counteradaptations.

   For theoretical accounts of co-evolution and arms races see Slatkin & Maynard Smith (1979) and Dawkins & Krebs (1979).

Chapter 13. Conclusion

The story of behaviour and adaptation that we have told in the last twelve chapters is inevitably too simple. All the 'ifs' and 'buts' of an impeccably cautious and impregnable account would have made the book twice as long and half as easy to understand. However we do not want to leave the impression that the ideas we have discussed are completely accepted by all evolutionary biologists. Far from it, even our basic assumptions are still very much disputed in the literature. In the first part of this chapter we look briefly at some of the controversies surrounding our two most important premises: selfish genes and the principle of optimality. Then in the second part of the chapter we will try to fill another gap by discussing how behavioural ecology relates to other approaches to the study of animal behaviour.

How plausible are our main premises?

SELFISH GENES

Our discussions of natural selection have always been along the following lines: 'Imagine a gene for such and such behaviour; when would it tend to spread in a population?'. As we saw in chapter 1 this approach does not imply that there are genes 'for' altruism, spite, long tails or whatever, but merely that there are some genetic differences between individuals which are correlated with the behaviour or structure in question.

But how plausible is the view that natural selection is a struggle between selfish genes rather than between individuals or groups? Obviously the field biologist sees *individuals* dying, surviving and reproducing; but the evolutionary consequence is that the frequencies of genes in the population change. Therefore the field biologist tends to think in terms of individual selection whilst the theorist thinks in terms of selfish genes. It can be very valuable, however, for the fieldworker to use selfish gene thinking in formulating ideas. This was apparent, for example, in chapter 9 where we saw that from the selfish gene viewpoint there is often no difference between parental care and sibling care, so hypotheses about the reasons for one and not the other occurring in a particular species must be framed in ecological rather than genetic terms. An exception to this rule, also clarified by selfish gene thinking, was in the social hymenoptera described in chapter 10 where females are genetically predisposed to help their sisters.

Thinking in terms of selfish genes can also help to reveal some subtle problems. For example the initial spread of a gene for altruism by kin selection is not as straightforward as it might at first sight appear. Imagine a brood of nestlings, all full siblings, one of which has a new mutant gene for altruistically giving some of its food to the others. Although the rest of the brood are kin of the altruist, they do not have the particular gene for altruistic food sharing. Therefore viewed from the point of view of a selfish gene, the altruistic individual does not make any genetic gain by giving up some of its food to boost the survival of its siblings. Altruism will be selectively disadvantageous even though the recipients are close kin. Only when the gene is common enough for most of the nestmates to share it will the stage be set for kin selection to favour its further spread or maintenance. How does the gene spread initially? One suggestion, made by Brian Charlesworth (1980), is that it may spread before it has any marked phenotypic effect or while only weakly influenced by selection. This is equivalent to the point made in chapter 1, that predictions of simple cost–benefit models of inclusive fitness may not correspond exactly to the predictions of more complete genetic models. For example genes may arise as recessive mutants and become dominant during evolution; this is what happened with the melanic gene of the peppered moth described in chapter 1. Another way to view the initial spread of an allele for altruism via kin selection is to think of kin selection as a special case of group selection. Within a kin group the altruist does less well than recipients, but groups with altruists do better than those without, so the trait spreads (see also next section).

Sometimes an analysis in terms of selfish genes may unravel a problem that appears puzzling when considered in the context of individuals maximising inclusive fitness. An example is the phenomenon of *segregation distortion*, in which an allele gets itself into a disproportionately large number of gametes. For a heterozygous parent, each allele is expected to occur in 50 per-cent of gametes, but segregation distorters somehow interfere with the process of gamete formation and increase their proportionate representation. In male *Drosophila* it is thought that chromosomes with segregation distorter alleles somehow cause sperms containing the homologous chromosome to become abnormal (for example they may have broken tails (Dawkins 1981)). This phenomenon can be understood within the framework of selfish genes competing for representation in future generations but is hard to explain in terms of maximising inclusive fitness. In fact since the effect of segregation distorter genes is to decrease fecundity (by killing off gametes), the interests of the selfish gene conflict with those of the individual.

An idea related to the concept of selfish genes has recently been proposed as an interpretation of the organisation of the genetic material itself. One of the recent discoveries of molecular biology is that the DNA in the cells of most higher organisms contains long repeated sequences. Some of these sequences are very simple repetitive units which are not transcribed when the genetic message is copied onto RNA for protein synthesis. (Other, more complex repeated sequences are used for coding for messsenger RNA.) It appears, then, that simple sequences do not function to provide messages for protein synthesis. A number of alternative functions has been suggested, for example they might act as structural elements. But none of these ideas is totally convincing. For example, they do not account for the fact that the amount of 'junk DNA' (as Orgel & Crick 1980 have dubbed it) varies from species to species, and for the extraordinary large proportion of the genome that is made up of junk in some animals (one species of crab has more than 60 per-cent junk DNA). The suggestion made by Orgel and Crick and by Doolittle and Sapienza (1980) is that the simple repeated sequences have no function for the cell. They exist purely for their own good. The proposal is that repeated sequences are bits of 'selfish DNA' analogous to selfish genes. Selfish genes spread through a population because they are good at surviving and reproducing (this is natural selection) and in a similar way selfish DNA spreads through the genome simply because it is good at making copies of itself. It does no good for the cell (in fact it presumably imposes a slight cost) but it spreads because it is an efficient replicator.

Selfish DNA is therefore rather like an intracellular parasite such as a virus. Bodies defend themselves against viruses (with the immune system) and one might expect the rest of the genome to resist in an analogous way the spread of selfish DNA. The extent to which selfish DNA spreads therefore probably reflects the outcome of an evolutionary race within the genome.

The parallel between selfish genes and selfish DNA provides a striking new way of looking at the structure organisation of the genome, but it should be borne in mind that the parallel is not exact. Selfish genes compete with their alleles to occupy the *same locus* in the genome, while selfish DNA spreads *laterally* through the genome.

GROUP SELECTION

At the beginning of the book we more or less dismissed group selection as a viable alternative to selection acting on individuals or selfish genes. We acknowledged that it could in principle work, but suggested that the conditions for group selection to be a powerful evolutionary force were not likely to be met in Nature very often. However this is not a universally accepted point of view. In particular D. S. Wilson (1980) has recently published a book in which he claims that group selection is a very major evolutionary force indeed and Leigh van Valen (1980), a distinguished evolutionary theorist, has heralded Wilson's book as a major breakthrough and a change of paradigm for evolutionary biologists. How should we treat this claim?

One point to bear in mind from the start is that Wilson's model is more subtle than the simple 'differential extinction of groups' model discussed in chapter 1.

The essential feature of Wilson's hypothesis is that populations are divided into groups (so-called 'trait groups') within which selection for or against altruistic traits (or any other trait for that matter) occurs. After selection has operated on the groups, the whole population mixes together before splitting up again into new groups for the next round of selection. There could be several ways in which individuals sort themselves into trait groups, but Wilson takes as the simplest case random assortment.

In Wilson's model the altruists are at a disadvantage *within* a trait group (because of their self sacrifice) but trait groups with altruists are more likely to contribute to the next generation than are trait groups with no altruists. If the population consisted of just one trait group, the altruistic gene would only spread if the fitness change for the actor (d) is greater than that for every other member of the population (r). But, and this is the crux of Wilson's argument, if the population is divided into many trait groups the effect of the altruistic act on the small proportion of non-altruists that happen to be in the same trait group as the altruist can be more or less ignored. For example if there were 100 bird flocks in a forest and one bird in one flock gave a warning call, the effect on the average fitness of all other birds in the forest would be quite small. Wilson concludes therefore that with trait groups, the conditions for an altruistic gene to spread are $d > 0$, instead of $d > r$ if there are no trait groups. (This means that any slight advantage to the altruist will cause the gene to spread regardless of the advantage obtained by recipients of the act.) Thus with selection acting on trait groups the altruistic gene is more likely to spread.

Wilson's model has been re-interpreted by Alan Grafen (in prep.). As shown in Box 13.1 Grafen points out that Wilson's result does not depend on the existence of trait groups but on population size. If the population is very large then the effect of an altruistic act on the small number of recipients nearby can be more or less ignored in calculations of average fitness. The issue has not been finally resolved, but at the moment it appears to be a semantic argument whether or not one refers to Wilson's model as one of group selection. What he has shown is that in a large population the effect of an altruistic act on neighbours is relatively unimportant. The emphasis on differential survival of trait groups seems to be unnecessary. The label 'group selection' could be said to be misleading because the effect Wilson describes is one that most evolutionary biologists would count as a straightforward example of individual selection. In favour of Wilson's model, it must be said that it demonstrates that the distinctions between individual, kin, and group selection discussed in chapter 1 are not clear cut.

Box 13.1 D. S. Wilson's model of group selection.
Wilson assumes that a potentially altruistic act can be characterised by its fitness change for the actor (d) and the average
fitness gain for every other member of the population (r). He
argues that in classical Darwinian selection the condition for
the spread of a gene for the act is $d > r$. In other words the act
has to benefit the actor more than any one else. However when
the population is divided into trait groups, Wilson argues, the
condition for the spread of the gene becomes $d > 0$. As long as
the act has some benefit for the actor, the benefit to others is
irrelevant. The argument is as follows:

(a) The classical condition. In a population of N individuals of
which A are altruists, the pay-offs to altruists and non-altruists
as a result of an altruistic act are as follows.

Altruist

$$d + \frac{(A-1)(N-1)r}{(N-1)} = d + (A-1)r$$

Gain from   Gain from acts
own act   of $(A-1)$ other
                   altruists each
                   giving $(N-1)r$ to
                   be shared among
                   $(N-1)$ recipients

Non-altruist

$$Ar$$

∴ *for altruism to spread*

$$d + (A-1)r > Ar$$
$$d > r \tag{1}$$

(b) Trait groups: The population is divided into T trait groups
each with N individuals. The total population is TN of which A
are altruists. The pay-offs from an altruitistic act are as follows:

Altruist

$$d + \frac{(A-1)(N-1)r}{TN-1}$$

Gain from   Gain from all other acts in trait group
own act   divided by total population

Non-altruist

$$\frac{A(N-1)r}{TN-1} = \frac{\text{Total paid out by altruists}}{\text{Population size}}$$

If there are many trait groups, TN is very large and so the
pay-off to an average non-altruist is very small. The condition
for the spread of altruism becomes:

$$d + \frac{(A-1)\,(N-1)r}{TN-1} > \frac{A(N-1)r}{TN-1}$$

The difference between these
is very small when TN is large

$$\therefore \approx d > 0 \qquad (2)$$

Therefore Wilson concludes that the division of a population into trait groups facilitates the spread of altruism, since condition (2) is less stringent than condition (1).

However, as Grafen points out, equation (2) would apply equally to a single population of TN individuals. The division into trait groups is not as important as population size.

OPTIMALITY MODELS AND ESSS

In nearly every chapter we have used the ideas of optimality and ESSs. An ESS is the equivalent of an optimal solution when the pay-offs are frequency dependent, so the advantages and limitations of the two kinds of model can be discussed together. In chapter 3 we encountered some of the criticisms which have been levelled at optimality arguments and some of the limitations in putting them into practice. To recap briefly there were three main points.

1 *The idea that animals are optimal cannot be tested.* As we saw in chapter 3 this criticism is based on a mistaken notion. The aim of using an optimality model is not to test whether animals are optimal, but to test whether the particular optimality criterion and constraints used in the model give a good description of the animal's behaviour.

2 *It is hard to tell why the animal's behaviour does not fit the predictions exactly.* Very often the simple models give an approximate but not exact description of the animal's behaviour. This could be because the model makes incorrect assumptions about constraints or goals, or because some component of cost has not been measured. There is no simple way of distinguishing between these possibilities.

3 *Animals are not well enough adapted to optimise.* The main rationale for using optimality and ESS models is the assumption that natural selection produces well adapted animals, the aim of the models being to find out how they are adapted. There are, however, at least three reasons why animals might not be well adapted.

(a) The physical or biological environment may fluctuate too rapidly for the animals to 'catch up' in their adaptations. For example the Atlantic gannet (*Sula bassana*) lays a clutch of one egg (Fig. 13.1), but when a second chick is experimentally added to the brood, both are reared without difficulty (Nelson 1964). This result appears to show that gannets are not well adapted to their environment, since natural selection should favour individuals that maximise their lifetime reproductive success (chapter 1). The reason that gannets are badly

Fig. 13.1 The Atlantic Gannet, here shown at the nest with a five week old chick, rears only one young per season, although brood manipulations suggest that it could raise two (Photo by J. B. Nelson).

adapted seems to be that the food supply has recently changed. Gannets now feed some of the time on fish offal discarded by fishing boats and this extra food allows them to rear more young than in former times, but selection has not yet had time to change the clutch size from one or two. This view is supported by the observation that the southern hemisphere gannet (*Sula capensis*) which feeds in waters relatively unaffected by fishermen can still raise only one chick to full fledging weight (Jarvis 1974).

(b) There may be insufficient genetic variation for new strategies to evolve. If the environment changes or if for some other reasons the optimal phenotype changes, animals can adapt to the new conditions only if there is genetic variation in the population. Although the issue is by no means resolved, it seems likely that in small populations the rate of evolution might well be limited by the rate at which new mutations arise (Maynard Smith 1975).

(c) There may be co-evolutionary arms races, as for example between predator and prey; if one side is ahead in the arms race, the other will appear to be poorly adapted to its environment, for example hosts that are killed or debilitated by pathogens and parasites.

Two more criticisms can be added to this list. One is that optimality and ESS models assume that genes exist, but contain no specific genetic mechanisms. For example, the ESS models of fighting strategies in chapter 5 did not allow for sexual reproduction and the mixing of genes that this entails. The attitude of the optimality or ESS theorist is 'think of the strategies and let the genes look after themselves'. The population geneticist on the other hand would like to know whether the models can really be couched in terms of genes and whether the rules of inheritance will allow the equilibria suggested by the models to actually develop. The answer to this question is largely unexploited at the moment. The second criticism is more specifically directed towards the optimality models used and tested in this book. The critical reader will have noticed that although we stressed the value of making quantitative predictions from optimality and ESS models, most of the tests of these predictions were qualitative. The animals were usually seen to do 'approximately the right thing' the dungflies in chapter 3, for example, copulated for 36 min instead of the predicted 41 min. Some might ask whether it is worth developing quantitative arguments if the tests are only qualitative. The answer is that tests are often qualitative simply because of limitations in the techniques used to carry out the tests. The value of quantitative predictions is still potentially just as great and what is needed is comparable technical developments in ways of testing the models. Once the quantitative predictions can be tested accurately, discrepancies between observed and predicted results help to tell us what is wrong with the models.

It is possible to carry on discussing the pros and cons of optimality models for a long time, but the strongest argument in their favour is that over and over again optimality arguments have helped us to understand adaptations. We have illustrated this point in the preceding chapters with behavioural examples; foraging, flock size, territory size and so on, but optimality arguments can equally well be used to understand adaptations at the physiological and biochemical level. For example the familiar 'herring bone' arrangement of the swimming muscles of many fish is not merely an incidental design feature. This arrangement allows the muscles to contract at a rate which maximises their power output (Alexander 1975). At the biochemical level the energy for muscle contraction is generated by oxidation of carbohydrates or fats via the Krebs cycle. It would be chemically feasible to carry out the oxidation by a more direct route, but the advantage of the cycle is that it maximises the net energy gain per molecule oxidised (Baldwin & Krebs 1981).

The comments discussed so far could apply equally well to simple optimality and to ESS models. Now let us briefly consider a case where the two kinds of model may lead to different interpretations of field data. Richard Dawkins and Jane Brockmann (1980) demonstrated that digger wasps adopt what appears at first sight to be a suboptimal strategy, but further analysis in terms of an ESS revealed a possible explanation as to why the particular strategy might be used. The behaviour in question was the persistence of fighting by female wasps disputing the ownership of a burrow (chapter 8). The question asked by Dawkins and Brockmann was 'how long should a female persist in a fight?'. They made the assumption that females are designed to maximise the benefit they obtain from the contest, which leads to the prediction that a female's persistence should be related to the value of the burrow. This in turn depends on the number of paralysed insects stored in the burrow. A nest containing four insects is ready for egg-laying and the winner of this prize saves herself days of digging and provisioning, while a burrow with only one insect still needs a lot of work before it is ready.

Contrary to expectation, Dawkins and Brockmann found that the persistence of females in a fight was not correlated with the total number of insects in the burrow but with the number put in by the loser. Since the loser is the one that determines the length of the fight (because the fight ends when the loser goes away), Dawkins and Brockmann concluded that wasps fight in proportion to their own past contribution rather than in proportion to the total value of the burrow. It is easy to see that this could lead to females giving up very quickly in a fight over a valuable burrow, just because the other wasp has done more of the provisioning, even though the pay-off goes to the winner regardless of who provisioned the burrow.

Dawkins and Brockmann's first reaction was that the wasps probably cannot 'tell' how many insects are in the burrow, and that the rule 'fight in proportion to own contribution' is a reasonable rule of thumb that approximates the optimal strategy, since there is usually some correlation between the total number and the number put in by each female. This is an example of how the optimal policy depends on assumptions about the constraints. A perfectly knowledgeable wasp should fight in proportion to total burrow contents, but when constrained by ignorance of the total number, the policy 'fight in proportion to own contribution' might be the best option.

However, now consider what would happen if all wasps were endowed with perfect knowledge. If both wasps had the same assessment of the value of the burrow and fought for a time proportional to this value, both would fight equally hard and give up at the same moment! Presumably chance factors would cause one wasp to persist fractionally longer than the other, and if these factors were truly random each wasp would have a 50:50 chance of winning after the long struggle. Imagine now a wasp that decided how long to fight by the toss of a coin and adopted the rule 'if heads give up at once, if tails persist indefinitely' (this is equivalent to the 'bourgeois' strategy in chapter 5). In a population of perfectly knowledgeable wasps this strategy would win half its fights and waste no time in lost contests. Therefore its net pay-off (subtracting total time wasted from the pay-off for winning) would be higher than that of the omniscient wasp. The message of this example is that 'perfect knowledge', which appeared at first sight to be an optimal strategy, is not an ESS. In general terms, whenever the pay-off depends on what others do, the question should be analysed as an ESS rather than a simple optimality problem.

Causal and functional explanations

[handwritten annotations: "how" above "Causal", "why" above "functional"]

Behavioural ecology is about functional explanations (the answers to 'why?' questions) of behaviour. As we emphasised in chapter 1 a great deal of misunderstanding can arise if functional and causal ('how?') explanations are confused. A simple illustration of this is an objection that is often raised to labels such as 'selfish', 'spiteful', 'sneak', 'transvestite', and 'rape' used by behavioural ecologists to describe the behaviours they observe. The objection is that the labels are too anthropomorphic and are loaded with the implication that animals are endowed with human-like motives. The answer to this objection is that the labels are used not to describe the causal mechanisms underlying the behaviour but to describe its functional consequences. When a male mallard 'rapes' a female he forcibly injects his sperm without going through the usual courtship ritual leading up to mating. The term rape is a good description of the consequences of the behaviour, but in using it the behavioural ecologist is not making any judgement about the motivational basis of the behaviour in mallards or any other animals.

Although it is important to be clear about the distinction between causal and functional explanations, it is equally valuable to realise that the two kinds of question are complementary and that asking 'why?' questions can often help to understand the answers to 'how?' questions, or vice versa. An example of how causal and functional explanations go hand in hand is illustrated in Fig. 13.2. Prairie dogs (*Cynomys ludovicianus*) are colonial and live in underground tunnels which may be up to 15 m long. The tunnels are usually simple U-shaped passages with an opening to the surface at either end. It has been known for a long time that prairie dogs build little mounds of soil around the two entrances of the tunnel. These mounds were considered to function either as lookout points or to protect the tunnel against floods. However a closer inspection revealed that the two ends of the burrow have different kinds of mounds. At one end there is a high steep sided 'crater' mound while at the other end there is a low rounded 'dome' (Fig. 13.2). If the mounds are simple lookouts or flood barriers, why should they be different shapes? The answer to this 'why?' question come from an understanding of how air is exchanged in the tunnel (Vogel, Ellington & Kilgore 1973). A prairie dog living in the long underground tunnel cannot survive without a regular supply of fresh air, and it appears that the mounds around the two entrances are designed to ensure a continuous flow of air through the tunnel. The crater mound is higher and has steeper sides than the dome; as a consequence, air is sucked out of the crater end of the tunnel and into the dome end.

The forces causing the air flow are viscous sucking and the Bernoulli effect. Viscous sucking refers to the fact that when moving air passes a region of stationary air, the still air is dragged along with the current. This effect is larger at the crater end because the crater is higher than the dome and so it is exposed to faster winds. The Bernoulli effect states that the pressure of a steadily moving fluid decreases when its velocity increases. The velocity of air above the crater is greater than above the mound and the crater has very still air inside because of its steep edges. The pressure drop between the inside and outside of the crater is therefore higher than in the case of the dome so the Bernoulli effect causes air to be sucked out of the crater end of the tunnel. Vogel *et al.* (1973) demonstrated by means of laboratory experiments with miniature model tunnels and by dropping smoke bombs down real tunnels in the field that the mound system is so effective that it causes the air in the tunnel to change once every ten minutes even in a very light breeze. The rate of air exchange is related to wind speed, but is unaffected by wind direction since the mounds are symmetrical. This second feature is important because wind directions is unpredictable in the prairie dog's natural habitat. This example illustrates that a functional question 'what are mounds for?' led to a detailed under-standing of a mechanism question 'how do prairie dogs get enough fresh air?'.

Fig. 13.2 A diagrammatic section of a prairie dog burrow. A typical burrow has two entrances, one with a low, round 'dome' at its entrance and the other with a taller, steeper-sided 'crater'. The different heights and shapes of the burrow entrances cause air to be sucked out of the crater end and therefore in through the dome (Vogel *et al.* 1973).

Let us now briefly consider how functional questions about behaviour might help in understanding three other types of question: mechanisms, ontogeny and learning.

(a) Mechanisms: motivation

One of the major questions for those interested in mechanisms is about the motivational basis of behaviour. For example in the case of feeding behaviour the question is 'how do internal physiological factors and external events combine to determine whether or not an animal is hungry?' A great deal is known about the physiological basis of hunger in terms of receptors in the brain, chemicals in the blood and stimuli from the digestive tract (Silverstone 1976). But as David McFarland (1976) has pointed out, in order to understand whether an animal will feed or do something else at a particular moment it is necessary to take into account functional considerations. If an animal is in a physiological condition which makes it extremely hungry, but the risk of going out to forage is very high because there are many predators around, the animal might postpone feeding until later on when the predators have gone to sleep or moved elsewhere. Another way of saying this is that the animal has built into its decision rules which involve calibrating its motivational state: the value of a certain physiological signal has to be adjusted according to the risks in the external environment. McFarland's line of argument leads to a very important link between causal and functional questions. The behavioural ecologist interested in decision rules (chapter 3) has to know about motivation since internal state is part of the animal's calculation of costs and benefits, while the behavioural physiologist interested in motivation needs to know about risks in the environment since these influence the animal's calibration of its internal state. There have been very few studies to date on how internal state and risk interact, but we saw one example in chapter 3: Milinski and Heller's study of hunger and predation risk in sticklebacks.

(b) Ontogeny: imprinting

Among the many important discoveries of the great ethologist Konrad Lorenz was the demonstration that newly hatched ducklings and goslings become attached to their mother very early in life as a result of rapid learning (known as imprinting). Young birds can be induced to follow and snuggle up to surrogate mothers such as balloons, cardboard boxes and scientists simply by exposing them to the object in question for a few hours during a sensitive period early in life. Learning what mother looks like is referred to as 'filial imprinting'. Even more striking is the finding that sexual preferences of ducks, geese and quite a few other birds can be modified by early experience

(Immelman 1975). When young birds are cross-fostered (raised by parents of a different species from their own) they will subsequently prefer to mate with members of their foster species rather than their own kind. The learned preference (referred to as 'sexual imprinting') seems to be very strong and persists even if the imprinted birds are kept for a long time with access only to members of their own species.

For more than 30 years after Lorenz first wrote about sexual imprinting there was no satisfactory functional explanation for it. Why should the choice of sexual partner be dependent on learning rather than a fixed genetic instruction? After all the fixed rule 'mate with your own species' would seem to be less risky than a preference based on learning.

In the past few years Patrick Bateson (1978, 1980) has provided evidence for a convincing functional explanation for sexual imprinting. Bateson's hypothesis is that imprinting allows a young animal to learn what its close kin look like (as well as its species characteristic) so that it can subsequently mate with members of its own species while avoiding mating with its own kin. Mating with kin is disadvantageous because it leads to 'inbreeding depression'—a decrease in the viability of offspring of close kin matings as compared with outbred offspring. (Although its genetic basis is not fully understood, inbreeding depression has been recorded in a number of species (see chapter 7). Perhaps the explanation is that inbred young are more likely to be homozygous for disadvantageous recessive genes.

Bateson has shown experimentally that Japanese quail (*Coturnix coturnix japonica*) prefer to mate with partners that differ slightly in plumage colour from their own parents. When normal brown males which had been reared in groups with other normal coloured quail were offered a choice between a brown and a mutant white female they preferred to mate with the former. But when faced with the choice of a familiar brown female (one with which they had been reared) and an unfamiliar one they preferred to approach and mate with the latter. In other words the males have the strongest preferences for *slightly* novel females. Although the familiar females were not sisters of the males, they were ones with which they had grown up, so Bateson suggests that as a result of early learning males avoid mating with members of their own brood. Normally this would prevent in breeding while the preference for familiar rather than very unfamiliar ensures that males choose the right species.

Another study in which avoidance of mating with close kin has been observed is on the Bewick's swan (*Cygnus columbianus*). These birds have characteristic facial markings (Fig. 13.3) and there is a tendency for members of the same family to have similar faces. Bateson, Lotwick and Scott (1980) found that mated pairs have facial patterns which are more different from one another than expected by chance, so it is possible that young birds learn their family characteristics and avoid incest by choosing to mate with partners whose faces clearly differ from their own family. Incest avoidance in man seems also to be based on early learning: as we saw in chapter 7, children reared together in an Israeli kibbutz avoid mating with each other although they were not close relatives.

(c) Learning: the matching law

The study of animal learning has traditionally been the province of comparative psychologists interested in mechanisms of learning and reinforcement, but some of the 'laws of learning' derived by psychologists can also be explained in functional terms.

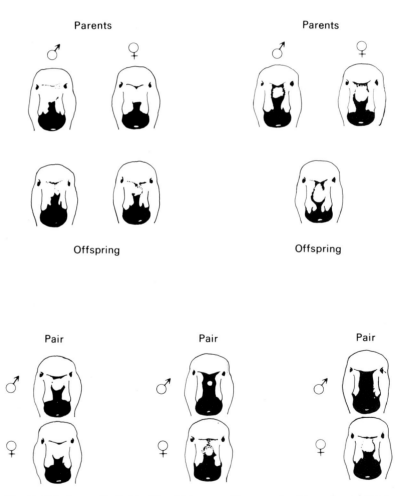

Fig. 13.3 The faces of individual Bewick's swans (*Cygnus columbianus bewickii*) are very distinctive. The top drawings show that offspring tend to resemble their parents. The lower drawings show that mates are often different in appearance. Therefore it appears that the facial pattern is inherited and the swans may actively outbreed (from Bateson, Lotwick & Scott 1980).

An example is the so-called 'matching law' (Herrnstein 1970). This law has been demonstrated in an experiment in which a hungry pigeon is trained to peck at a key to obtain food and then offered a choice of two keys. A peck on either key sometimes yields brief access to food and the two keys provide rewards independently and at different rates.

The frequency with which a reward is offered is referred to as the reinforcement schedule. A commonly used schedule in simultaneous choice experiments is the *variable interval* schedule. In this schedule the first peck to a particular key after an average of t seconds since the last reward from that key will produce food. The average value of t determines the overall reward rate: a large value of t produces a low reward rate and vice versa. Usually the two keys have different values of t and the values have a random distribution about their means so that the animal can predict only approximately when the next reward is due. This schedule is analogous to the natural situation in which an animal temporarily depletes its food source. For example, when a hummingbird drinks the nectar from a flower, it is not likely to get another reward from the same flower until after a certain time has elapsed. In order to maximise its overall reward rate the bird would have to return after sufficient time without waiting too long and thereby missing the opportunity to use a good source of food.

In choice experiments with variable interval schedules pigeons and other animals have been found to allocate their responses to the two keys in proportion to the ratio of reinforcements obtained from the keys in the past. This is referred to as *matching*. The result can be summarised in the following equation:

$$\frac{\text{responses to key 1}}{\text{responses to key 2}} = k \left\{ \frac{\text{rewards obtained from 1}}{\text{rewards obtained from 2}} \right\}^{x}$$

where k is a constant indicating any bias to one key or the other ($k = 1$ means no bias) and x describes the relationship between response ratio and reward ratio. If $x = 1$ the relationship is a simple linear one, and this is the pattern usually shown by pigeons.

This generalisation was originally derived inductively and was used as a description of the *mechanism* relating learning and reinforcement. However it can also be viewed from a functional standpoint, in which case it turns out that if average reward rate is maximised, the animal will approximate matching (Staddon 1980). The matching rule therefore describes the pattern of responding that would have been predicted from optimal foraging models of the sort described in chapter 3.

Another way in which functional questions can contribute to the study of learning is by making predictions about the extent to which animals should expend effort in learning about their environment. For example a foraging animal in a very unpredictable environment would benefit less from time spent in exploration than would one in a predictable environment in which past experience would be useful in predicting good feeding places for the future. This idea is supported by an experiment in which captive great tits spent more time exploring the environment for good feeding sites when they were accustomed to long term stability of good feeding sites than when the quality of sites fluctuated on a short term basis (Kacelnik 1979b).

A final comment

What we have described as behavioural ecology in this book is the present day equivalent of natural history. It stands in a lineage which gradually developed from detailed descriptions of animal behaviour by naturalists such as Gilbert White and Henri Fabre to experimental studies of natural history by Tinbergen and others. At the moment there is something of a bandwagon in natural history for inventing functional explanations of behaviour, which gives behavioural ecology and sociobiology a bad name. We have tried to avoid this as much as possible (without complete success!) and instead we have emphasised the idea of making testable predictions about adaptation. To illustrate how this approach has developed from studies of natural history let us construct a hypothetical lineage of studies of mating behaviour in dungflies.

A few hundred years ago naturalists would have been satisfied to discover that when two dungflies were seen together, one riding on the other's back, the top one was a male and the one underneath was a female and that the two were mating. A hundred years ago Darwin realised that males in general compete for females. A description of the natural history of dungfly mating at this stage would have included reference to the fact that males are larger than females and that this may be a result of sexual selection. Ten years ago evolutionary biologists would have stressed the idea that males ride on the backs of females not only to inject sperm, but also for some time after copulation while the female lays her eggs. By guarding the female in this way the male guarantees that his sperm are not displaced by those of another male. In the last few years behavioural ecologists have started to try and explain why it is that the male rides on the female's back for 40 min and not 10, 20 or 60. In developing a theory to explain why, it has become apparent that the same kind of analysis can be used for bumblebees sucking nectar out of flowers, parents investing in their offspring and many other problems. This gradual reductionist progression from broad description to detailed quantitative analysis and simple generalisations is one of the major themes of development of the natural history lineage. But at the same time natural historians are interested in differences between species. Perhaps in the next few years we will see the same quantitative approach used to explain why different species behave in different ways.

Summary

This chapter is in two parts. The first section tries to assess the value and the limitations of the selfish gene and optimality views of evolution. Group selection as an alternative to individual selection is reassessed in the light of D. S. Wilson's work. The value of optimality arguments can be illustrated by studies of adaptations at the behavioural, physiological and biochemical levels.

In the second part of the chapter, the distinction between 'how' and 'why' questions is stressed. Also we try to show how different kinds of questions (function, causation, and ontogeny) should go hand in hand in studies of behaviour.

Further reading

Maynard Smith (1977b) presents a very clear and thought-provoking summary of unsolved problems in evolutionary theory of relevance to behavioural ecologists using optimality arguments. Particularly valuable is his discussion of what limits the rate of evolution.

Gould (1980) is also a critical and stimulating review of neo-Darwinian evolution. He argues that microevolutionary changes of the type we have discussed in this book may be caused by different mechanisms from those leading to macroevolutionary changes (evolution and extinction of species).

Pulliam & Dunford (1980) attempt to produce an account of the adaptive value of learning and culture. This is an important new area of behavioural ecology.

Dawkins (1981) is a detailed development of some of the ideas first presented in *The Selfish Gene*.

References

Abele, L.G. & Gilchrist, S. 1977. Homosexual rape and sexual selection in acanthocephalan worms. *Science, N.Y.* **197**, 81–3.

Alcock, J. 1979. *Animal Behaviour: an Evolutionary Approach*. 2nd edition. Sinauer, Sunderland, Massachusetts.

Alcock, J., Jones, C.E. & Buchmann, S.L. 1977. Male mating strategies in the bee *Centris pallida*, Fox (Anthophoridae: Hymenoptera). *Amer. Natur.* **111**, 145–55.

Alexander, R.D. 1974. The evolution of social behavior. *Ann. Rev. Ecol. Syst.* **5**, 325–83.

Alexander, R.D. 1975. Natural selection and specialized chorusing behaviour in acoustical insects. In Pimentel, D. (ed.). *Insects Science and Society*, pp. 35–77. Academic Press, New York.

Alexander, R.D. & Borgia, G. 1979. On the origin and basis of the male–female phenomenom. In M.S. Blum & N.A. Blum (eds). *Sexual Selection and Reproductive Competition in Insects*. Academic Press, New York.

Alexander, R.D. & Sherman, P.W. 1977. Local mate competition and parental investment in social insects. *Science, N.Y.* **196**, 494–500.

Alexander, R.D. & Tinkle, D.W. (eds). 1981. *Natural Selection and Social Behaviour: Recent Research and New Theory*. Chiron Press, New York.

Alexander, R.McN. 1975. *The Chordates*. Cambridge University Press, Cambridge.

Altmann, S.A. & Altmann, J. 1970. *Baboon Ecology*. University of Chicago Press, Chicago.

Andersson, M. 1980. Why are there so many threat displays? *J. theor. Biol.* **86**, 773–81.

Andersson, M. & Wicklund, C.G. 1978. Clumping versus spacing out: experiments on nest predation in fieldfares (*Turdus pilaris*). *Anim. Behav.* **26**, 1207–12.

Aoki, S. 1977.*Colophina clematis* (Homoptera: Pemphigidae), an aphid species with 'soldiers'. *Kontyû, Tokyo*, **45**, 276–82.

Argyle, M. 1972. Non-verbal communication in human social interaction. In R.A. Hinde (ed.). *Non-Verbal Communication*, pp. 243–69. Cambridge University Press, Cambridge.

Arnold, S.J. 1976. Sexual behaviour, sexual interference and sexual defence in the salamanders *Ambystoma maculatum*, *A. tigrinum* and *Plethodon jordani*. *Z. Tierpsychol.* **42**, 247–300.

Baldwin, J.E. & Krebs, H.A. 1981. The evolution of metabolic cycles. *Nature, Lond.* **291**, 381–382.

Barnard, C.J. 1980. Flock feeding and time budgets in the house sparrow, *Passer domesticus*, L. *Anim. Behav.* **28**, 295–309.

Bateman, A.J. 1948. Intra-sexual selection in *Drosophila*. *Heredity* **2**, 349–68.

Bateson, P.P.G. 1978. Sexual imprinting and optimal outbreeding. *Nature, Lond.* **273**, 659–60.

Bateson, P.P.G. 1980. Optimal outbreeding and the development of sexual preferences in Japanese quail. *Z. Tierpsychol.* **53**, 231–44.

Bateson, P.P.G., Lotwick, W. & Scott, D.K. 1980. Similarities between the faces of parents and offspring in Bewick's swans and the differences between mates. *J. Zool. Lond.* **191**, 61–74.

Baylis, J.R. 1981. The evolution of parental care in freshwater fishes with reference to Darwin's rule of male sexual selection (in prep).

Bell, G. 1978. The evolution of anisogamy. *J. theor. Biol.* **73**, 247–70.

Belovsky, G.E. 1978. Diet optimization in a generalist herbivore: the moose. *Theoret. Pop. Biol.* **14**, 105–34.

Benzer, S. 1973. Genetic dissection of behavior. *Sci. Amer.* **229** (12), 24–37.

Bertram, B.C.R. 1975. Social factors influencing reproduction in wild lions. *J. Zool. Lond.* **177**, 463–82.

Bertram, B.C.R. 1976. Kin selection in lions and in evolution. In P.P.G. Bateson & R.A. Hinde (eds). *Growing Points in Ethology*, pp. 281–301. Cambridge University Press, Cambridge.

Bertram, B.C.R. 1978. Living in groups: predators and prey. In J.R. Krebs & N.B. Davies (eds). *Behavioural Ecology: an Evolutionary Approach*, pp. 64–96. Blackwell Scientific Publications, Oxford.

Bertram, B.C.R. 1979. Ostriches recognise their own eggs and discard others. *Nature, Lond.* **279**, 233–4.

Bertram, B.C.R. 1980. Vigilance and group size in ostriches. *Anim. Behav.* **28**, 278–86.

Birkhead, T.R. 1977. The effect of habitat and density on breeding success in common guillemots, *Uria aalge. J. Anim. Ecol.* **46**, 751–64.

Birkhead, T.R. 1979. Mate guarding in the magpie *Pica pica. Anim. Behav.* **27**, 866–74.

Birkhead, T.R. & Clarkson, K. 1980. Mate selection and precopulatory guarding in *Gammarus pulex. Z. Tierpsychol.* **52**, 365–80.

Blum, M.S. & Blum N.A. (eds) 1979. *Sexual Selection and Reproductive Competition in Insects*. Academic Press, New York.

Blurton Jones, N.G. 1968. Observations and experiments on causation of threat displays of the great tit, *Parus major. Anim. Behav. Monogr.* **1**, 75–158.

Borgia, G. 1979. Sexual selection and the evolution of mating systems. In M.S. Blum & N.A. Blum (eds) *Sexual Selection and Reproductive Competition in Insects*, pp. 19–80. Academic Press, New York.

Bowman, R.I. 1981. Adaptive morphology of song dialects in Darwin's finches. *Proc. XVII Int. Ornithol. Congr. (Berlin).* (in press).

Bradbury, J.W. 1977. Lek mating behaviour in the hammer-headed bat. *Z. Tierpsychol.* **45**, 225–55.

Bradbury, J.W. & Vehrencamp, S.L. 1977. Social organization and foraging in emballonurid bats. III. Mating systems. *Behav. Ecol. Sociobiol.* **2**, 1–17.

Bray, O.E., Kennelly, J.J. & Guarino, J.L. 1975 Fertility of eggs produced on territories of vasectomized red-winged blackbirds. *Wilson Bull.* **87**, 187–95.

Breder, C.M. & Rosen, D.E. 1966. *Modes of Reproduction in Fishes*. Natural History Press, New York.

Brockmann, H.J. & Dawkins, R. 1979. Joint nesting in a digger wasp as an evolutionarily stable preadaptation to social life. *Behaviour* **71**, 203–45.

Brockmann, H.J, Grafen, A. & Dawkins, R. 1979. Evolutionarily stable nesting strategy in a digger wasp. *J. theor. Biol.* **77**, 473–96.

Brooke, M. de L. 1978. Some factors affecting the laying date, incubation and breeding success of the manx shearwater, *Puffinus puffinus. J. Anim. Ecol.* **47**, 477–95.

Brown, J.L. 1964. The evolution of diversity in avian teritorial systems. *Wilson Bull.* **76**, 160–9.

Brown, J.L. 1969. The buffer effect and productivity in tit populations. *Amer. Natur.* **103**, 347–54.

Brown, J.L. 1978. Avian communal breeding systems. *Ann. Rev. Ecol. Syst.* **9**, 123–55.

Brown, J.L. & Brown, E.R. 1981. Kin selection and individual selection in babblers. In R.D. Alexander & D.W. Tinkle (eds). *Natural Selection and Social Behaviour: Recent Research and New Theory*. Chiron Press, New York.

Brown, J.L., Dow, D.D., Brown, E.R. & Brown, S.D. 1978. Effects of helpers on feeding of nestlings in the grey-crowned babbler, *Pomatostomus temporalis. Behav. Ecol. Sociobiol.* **4**, 43–60.

Busnel. R.G. & Klasse, A. 1976. *Whistled Languages*. Springer-Verlag, Berlin.

Bygott, J.D. Bertram, B.C.R. & Hanby, J.P. 1979. Male lions in large coalitions gain reproductive advantage. *Nature, Lond.* **282**, 839–41.

Cade, W. 1979. The evolution of alternative male reproductive strategies in field crickets. In M. Blum & N.A. Blum (eds). *Sexual Selection and Reproductive Competition in Insects.* pp. 343–79. Academic Press, London.

Calvert, W.H., Hedrick, L.E. & Brower, L.P. 1979. Mortality of the monarch butterfly, *Danaus plexippus:* avian predation at five overwintering sites in Mexico. *Science, N.Y.* **204**, 847–51.

Capranica, R.R., Frishkopf, L.S. & Nevo, E. 1973. Encoding of geographic dialects in the auditory system of the cricket frog. *Science, N.Y.* **182**, 1272–5.

Caraco, T. 1979a. Time budgeting and group size: a theory. *Ecology* **60**, 611–7.

Caraco, T. 1979b. Time budgeting and group size: a test of theory. *Ecology* **60**, 618–27.

Caraco, T. Martindale, S, & Pulliam H.R. 1980. Flocking: advantages and disadvantages. *Nature* **285**, 400–1.

Caraco, T., Martindale, S. & Whitham, T.S. 1980. An empirical demonstration of risk-sensitive foraging preferences. *Anim. Behav.* **28**, 820–30.

Caraco. T. & Wolf, L.L. 1975. Ecological determinants of group sizes of foraging lions. *Amer. Natur.* **109**, 343–52.

Carayon, J. 1974. Insémination traumatique hétérosexuelle et homosexuelle chez *Xylocoris maculipennis* (Hem. Anthocoridae). *C. R. Acad. Sc. Paris, D.* **278**, 2803–6.

Carpenter, C.R. 1954. Tentative generalisations on the grouping behaviour of non-human primates. *Human Biol.* **26**, 269–76.

Carpenter, F.L. & MacMillen, F.E. 1976. Threshold model of feeding territoriality and test with a Hawaiian honey creeper. *Science, N.Y.* **194**, 639–42.

Caryl. P. 1979. Communication by agonistic displays: what can games theory contribute to ethology? *Behaviour* **68**, 136–69.

Caryl, P. 1980. Escalated fighting and the war of nerves: games theory and animal combat. In P.P.G. Bateson & P.H. Klopfer (eds). *Perspectives in Ethology.* Plenum Press, New York.

Catchpole, C.K. 1979. *Vocal Communication in Birds.* Edward Arnold, London.

Catchpole, C.K. 1980. Sexual selection and the evolution of complex songs among European warblers of the genus *Acrocephalus. Behaviour* **74**, 149–66.

Chappuis, C. 1971. Un example de l'influence du milieu sur les émissions vocales des oiseaux: l'evolution des chants en forêt équitoriale. *Terre Vie* **25**, 183–202.

Charlesworth, B. 1980. Models of kin selection. In H. Markl, (ed.). *Evolution of Social Behavior: Hypotheses and Empirical Tests.* pp. 11–26. Dahlem konferenzen. Verlag Chemie, Weinheim.

Charnov, E.L. 1978. Evolution of eusocial behavior: offspring choice or parental parasitism? *J. theor. Biol.* **75**, 451–66.

Charnov, E.L. 1979. Natural selection and sex change in pandalid shrimp: test of a life history theory. *Amer. Natur.* **113**, 715–34.

Charnov, E.L. 1981. Kin selection and helpers at the nest: effects of paternity and biparental care. *Anim. Behav.* (in press).

Charnov, E.L., Orians, G.H. & Hyatt, K. 1976. The ecological implications of resource depression. *Amer. Natur.* **110**, 247–59.

Clarke, T.A. 1970. Territorial behavior and population dynamics of a pomacentrid fish, the garibaldi, *Hypsypops rubicunda.* (Pomacentridae). *Ecol. Monogr.* **40**, 180–212.

Clutton-Brock, T.H. 1975. Feeding behaviour of red colobus and black and white colobus in East Africa. *Folia primatol.* **23**, 165–207.

Clutton-Brock, T.H. & Albon, S.D. 1979. The roaring of red deer and the evolution of honest advertisement. *Behaviour* **69**, 145–70.

Clutton-Brock, T.H., Albon, S.D. Gibson, R.M. & Guinness, F.E. 1979. The logical stag: adaptive aspects of fighting in red deer (*Cervus elaphus* L.). *Anim. Behav.* **27**, 211–25.

Clutton-Brock, T.H. & Harvey, P.H. 1977. Primate ecology and social organisation. *J. Zool. Lond.* **183**, 1–39.

Clutton-Brock, T.H. & Harvey, P.H. 1979. Comparison and adaptation. *Proc. R. Soc. Lond. B.* **205**, 547–65.

Clutton-Brock, T.H. & Harvey, P.H. 1980. Primates, brains and ecology. *J. Zool. Lond.* **190**, 309–23.

Corbet, P.S. 1962. *A Biology of Dragonflies.* Witherby, London.

Coulson, J.C. 1966. The influence of the pairbond and age on the breeding biology of the kittiwake gull, *Rissa tridactyla. J. Anim. Ecol.* **35**, 269–79.

Cowie, R.J. 1977. Optimal foraging in great tits, *Parus major. Nature, Lond.* **268**, 137–39.

Cox, C.R. & Le Boeuf, B.J. 1977. Female incitation of male competition: a mechanism of mate selection. *Amer. Natur.* **111**, 317–35.

Cronin, E.W. & Sherman, P.W. 1977. A resource-based mating system: the orange-rumped honey guide. *Living Bird* **15**, 5–32.

Crook, J.H. 1964. The evolution of social organisation and visual communication in the weaver birds (Ploceinae). *Behaviour Suppl.*, **10**, 1–178.

Crook. J.H. & Gartlan, J.S. 1966. Evolution of primate societies. *Nature, Lond.* **210**, 1200–3.

Cullen, J.M. 1966. Reduction of ambiguity through ritualization. *Phil. Trans, Roy. Soc. B.* **251**, 363–74.

Cullen, J.M. 1972. Some principles of animal communication. In R.A. Hinde (ed.) *Non-Verbal Communication*, pp 101–22. Cambridge University Press, Cambridge.

Daly, M. 1979. Why don't male mammals lactate? *J. theor. Biol.* **78**, 325–45.

Darwin, C. 1859. *On the Origin of Species.* Murray, London.

Darwin, C. 1871. *The Descent of Man and Selection in Relation to Sex.* Murray, London.

Davies, N.B. 1978a. Territorial defence in the speckled wood butterfly (*Pararge aegeria*): the resident always wins. *Anim. Behav.* **26**, 138–47.

Davies, N.B. 1978b. Parental meanness and offspring independence: an experiment with hand-reared great tits, *Parus major. Ibis* **120**, 509–14.

Davies, N.B. & Halliday, T.R. 1978. Deep croaks and fighting assessment in toads *Bufo bufo. Nature, Lond.* **274**, 683–5.

Davies, N.B. & Houston, A.I. 1981. Owners and satellites: the economics of territory defence in the pied wagtail, *Motacilla alba. J. Anim. Ecol.* **50**, 157–80.

Dawkins, R. 1976. *The Selfish Gene.* Oxford University Press, Oxford.

Dawkins, R. 1978. Replicator selection and the extended phenotype. *Z. Tierpsychol.* **47**, 61–76.

Dawkins, R. 1979. Twelve misunderstandings of kin selection. *Z. Tierpsychol.* **51**, 184–200.

Dawkins, R. 1980. Good strategy or evolutionarily stable strategy? In G.W. Barlow & J. Silverberg. (eds). *Sociobiology: Beyond Nature/Nurture*, pp. 331–67. West-view Press, Boulder, Colorado.

Dawkins, R. 1981. *The Extended Phenotype.* W.H. Freeman, Oxford.

Dawkins, R. & Brockmann, H.J. 1980. Do digger wasps commit the Concorde fallacy? *Anim. Behav.* **28**, 892–6.

Dawkins, R. & Carlisle, T.R. 1976. Parental investment, mate desertion and a fallacy. *Nature, Lond.* **262**, 131–3.

Dawkins, R. & Krebs, J.R. 1978. Animal signals: information or manipulation? In J.R. Krebs & N.B. Davies (eds). *Behavioural Ecology: an Evolutionary Approach*, pp. 282–309. Blackwell Scientific Publications, Oxford.

Dawkins, R. & Krebs, J.R. 1979. Arms races between and within species. *Proc. R. Soc. Lond. B.* **205**, 489–511.

De Groot, P. 1980. A study of the acquisition of information concerning resources by individuals in small groups of red-billed weaver birds *Quelea quelea*. Ph.D. thesis, University of Bristol.

De Vore, I. (ed.). 1965. *Primate Behaviour: field studies of monkeys and apes.* Holt, Rinehart & Winston, New York.

Doolittle, W.F. & Sapienza C. 1980. Selfish genes, the phenotype paradigm and genome evolution. *Nature* **284**, 601–3.

Downhower, J.F. & Armitage, K.B. 1971. The yellow-bellied marmot and the evolution of polygamy. *Amer. Natur.* **105**, 355–70.

Dunbar, R.I.M. 1983. The logic of intraspecific variations in mating strategy. In P.P.G. Eateson & P. Klopfer (eds), *Perspectives in Ethology Vol. 5*. Plenum Press, New York. (in press).

Dunbar, R.I.M. & Dunbar, E.P. 1975. Social dynamics of gelada baboons. *Contr. Primatol.* **6**.

Duncan, P. & Vigne, N. 1979. The effect of group size in horses on the rate of attacks by blood-sucking flies. *Anim. Behav.* **27**, 623–5.

Dybas, H.S. & Lloyd, M. 1974. The habitats of 17 year periodical cicadas (Homoptera: Cicadidae: *Magicicada* spp.). *Ecol. Monogr.* **44**, 279–324.

Eberhard, W.G. 1972. Altruistic behaviour in a sphecid wasp: support for kin-selection theory. *Science* **175**, 1390–1.

Eberhard, W.G. 1979. The functions of horns in *Podischnus agenor* (Dynastinae) and other beetles. In M.S. Blum & N.A. Blum (eds). *Sexual Selection and Reproductive Competition in Insects*, pp. 231–58. Academic Press, London.

Edmunds, M. 1974. *Defence in Animals: a Survey of Anti-predator Defences*. Longman, Harlow.

Elner, R.W. & Hughes, R.N. 1978. Energy maximization in the diet of the shore crab, *Carcinus maenas*. *J. Anim. Ecol.* **47**, 103–16.

Emlen, S.T. 1978. The evolution of co-operative breeding in birds. In J.R. Krebs & N.B. Davies (eds). *Behavioural Ecology: an Evolutionary Approach*, pp. 245–81. Blackwell Scientific Publications, Oxford.

Emlen, S.T. & Oring, L.W. 1977. Ecology, sexual selection and the evolution of mating systems. *Science, N.Y.* **197**, 215–23.

Erickson, C.J. & Zenone, P.G. 1976. Courtship differences in male ring doves: avoidance of cuckoldry? *Science, N.Y.* **192**, 1353–4.

Evans, H.E. 1977. Extrinsic and intrinsic factors in the evolution of insect sociality. *BioScience* **27**, 613–7.

Fischer, E.A. 1980. The relationship between mating system and simultaneous hermaphroditism in the coral reef fish, *Hypoplectus nigricans*. *Anim. Behav.* **28**, 620–33.

Fisher, R.A. 1930. *The Genetical Theory of Natural Selection*. Clarendon Press, Oxford.

Fretwell, S.D. 1972. *Populations in a Seasonal Environment*. Princeton, Princeton University Press.

Fretwell, S.D. & Lucas, H.L. 1970. On territorial behaviour and other factors influencing habitat distribution in birds. *Acta Biotheoretica* **19**, 16–36.

Fricke, H.W. 1979. Mating system, resource defence and sex change in the anemonefish, *Amphiprion akallopisos*. *Z. Tierpsychol.* **50**, 313–26.

Fricke, H. & Fricke, S. 1977. Monogamy and sex change by aggressive dominance in coral reef fish. *Nature, Lond.* **266**, 830–2.

Frisch, K. von. 1967. *The Dance Language and Orientation of Bees*. Belknap Press, Cambridge, MA.

Gadgil, M. 1972. Male dimorphism as a consequence of sexual selection. *Amer, Natur.* **106**, 574–80.

Gaston, A.J. 1978. The evolution of group territorial behaviour and cooperative breeding. *Amer. Natur.* **112**, 1091–100.

Geist, V. 1966. The evolution of horn-like organs. *Behaviour* **27**, 175–213.

Geist, V. 1974. On fighting strategies in animal conflict. *Nature, Lond.* **250**, 354.

Ghiselin, M.T. 1969. The evolution of hermaphroditism among animals. *Q. Rev. Biol.* **44**, 189–208.

Gilbert, L.E. 1976. Postmating female odor in *Heliconius* butterflies: a male contributed antiaphrodisiac? *Science, N.Y.* **193**, 419–20.

Gilbert, L.E, & Raven, P.H. (eds). 1975. *Coevolution of animals and plants*. University of Texas Press, Austin & London.

Gill, F.B. & Wolf, L.L. 1975. Economics of feeding territoriality in the golden-winged sunbird. *Ecology* **56**, 333–45.

Goodall, J. 1968. The behaviour of free-living chimpanzees in the Gombe Stream Reserve. *Anim. Behav. Monogr.* **1**, 165–301.

Goss-Custard, J.D. 1970. Feeding dispersion in some overwintering wading birds. In J.H. Crook (ed.) *Social Behaviour in Birds and Mammals*. pp. 3–34. Academic Press, London.

Goss-Custard, J.D. 1976. Variation in the dispersion of redshank (*Tringa totanus*) on their winter feeding grounds. *Ibis* **118**, 257–63.

Gould, S.J. 1966. Allometry and size in ontogeny and phylogeny. *Biol. Rev.* **41**, 587–640.

Gould, S.J. 1978. *Ever Since Darwin: Reflections in Natural History*. Andre Deutsch, London.

Gould. S.J. 1980. Is a new and general theory of evolution emerging? *Paleobiology* **6**, 119–30.

Gould, S.J. & Lewontin, R.C. 1979. The spandrels of San Marco and the Panglossian paradigm: a critique of the adaptationist programme. *Proc. R. Soc. Lond. B.* **205**, 581–98.

Greenberg, L. 1979. Genetic component of kin recognition in primitively social bee. *Science, N.Y.* **206**, 1095–7.

Greenwood, P.J. 1980. Mating systems, philopatry and dispersal in birds and mammals. *Anim. Behav.* **28**, 1140–62.

Greenwood, P.J., Harvey, P.H. & Perrins, C.M. 1978. Inbreeding and dispersal in the great tit. *Nature, Lond*, **271**, 52–4.

Hailman, J.P. 1977. *Optical Signals*. Indiana University Press, Bloomington.

Haldane, J.B.S. 1953. Animal populations and their regulation. *Penguin Modern Biology* **15**, 9–24.

Halliday, T.R. 1978. Sexual selection and mate choice. In J.R. Krebs & N.B. Davies, (eds). *Behavioural Ecology: an Evolutionary Approach*, pp. 180–213. Blackwell Scientific Publications, Oxford.

Hamilton, W.D. 1964. The genetical evolution of social behaviour. I, II. *J. theor. Biol.* **7**, 1–52.

Hamilton, W.D. 1967. Extraordinary sex ratios. *Science, N.Y.* **156**, 477–88.

Hamilton, W.D. 1971. Geometry for the selfish herd. *J. theor. Biol.* **31**, 295–311.

Hamilton, W.D. 1972. Altruism and related phenomena, mainly in social insects. *Ann. Rev. Ecol. Syst.* **3**, 193–232.

Hamilton, W.D. 1979. Wingless and fighting males in fig wasps and other insects. In M.S. Blum & N.A. Blum (eds). *Sexual Selection and Reproductive Competition in Insects*, pp. 167–220. Academic Press, London.

Harvey, P.H., Kavanagh, M. & Clutton-Brock, T.H. 1978. Sexual dimorphism in primate teeth. *J. Zool. Lond.* **186**, 475–86.

Harvey, P.H. & Mace, G.M. 1981. Comparison between taxa and adaptive trends: problems of methodology. In Bertram, B.C.R. *et al.* (eds). *Current Problems in Sociobiology*. Cambridge University Press, Cambridge.

Heinrich, B. 1979. *Bumblebee Economics*. Harvard University Press, Cambridge, Mass.

Heller, R. & Milinski, M. 1979. Optimal foraging of sticklebacks on swarming prey. *Anim. Behav.* **27**, 1127–41.

Henwood. K. & Fabrick, A. 1979. A quantitative analysis of the dawn chorus: temporal selection for communicatory optimisation. *Amer. Natur.* **114**, 260–74.

Herrnstein, R.J. 1970. On the law of effect. *J. Exp. Anal. Behav.* **13**, 243–66.

Hinde. R.A. 1956. The biological significance of the territories of birds. *Ibis* **98**, 340–69.

Hinde, R.A. 1970. *Animal Behaviour: A Synthesis of Ethology and Comparative Psychology*. 2nd edition. McGraw-Hill, New York.

Hinde, R.A. 1975. The concept of function. In G. Baerends, C. Beer & A. Manning (eds). *Function and Evolution of Behaviour*, pp. 3–15. Clarendon Press, Oxford.

Hinde, R.A. 1981. Animal signals: ethological and games-theory approaches are not incompatible. *Anim. Behav.* (in press).

Hogan-Warburg, A.J. 1966. Social behaviour of the ruff, *Philomachus pugnax* (L.). *Ardea* **54**, 109–229.

Holldöbler, B. 1977. Communication in social Hymenoptera. In T.A. Sebeok (ed.). *How Animals Communicate*, pp. 418–71. Indiana University Press, Blooming-ton & London.

Holm, C.H. 1973. Breeding sex ratios, territoriality and reproductive success in the red-winged blackbird (*Agelaius phoeniceus*). *Ecology* **54**, 356–65.

Hoogland, J.L. 1979a. The effect of colony size on individual alertness of prairie dogs (Sciuridae: *Cynomys* spp.) *Anim. Behav.* **27**, 394–407.

Hoogland, J.L. 1979b. Aggression, ectoparasitism and other possible costs of prairie dog (Sciuridae: *Cynomys* spp.) coloniality. *Behaviour* **69**, 1–35.

Howard, R.D. 1978a. Factors influencing early embryo mortality in bullfrogs. *Ecology* **59**, 789–98.

Howard, R.D. 1978b. The evolution of mating strategies in bullfrogs, *Rana catesbia-na*. *Evolution* **32**, 850–71.

Howard, R.D. 1979. Big bullfrogs in a little pond. *Nat. Hist. Mag.* **88**, 30–6.

Hrdy, S.B. 1977. *The Langurs of Abu: Female and Male Strategies of Reproduction*. Harvard University Press, Cambridge, Mass.

Hunter, M.L. & Krebs, J.R. 1979. Geographical variation in the song of the great tit (*Parus major*) in relation to ecological factors. *J. Anim. Ecol.* **48**, 759–85.

Huxley, J.S. 1966. A discussion of ritualisation of behaviour in animals and man: Introduction. *Phil. Trans. R. Soc. B*, **251**, 247–71.

Hyatt, G.W. & Salmon, M. 1978. Combat in the fiddler crabs *Uca pugilator* and *U. pugnax*: a quantitative analysis. *Behaviour* **65**, 182–211.

Immelmann, K. 1975. Ecological significance of imprinting and early learning. *Ann. Rev. Ecol. Syst.* **6**, 15–37.

Janzen, D.H. 1979. Why fruit rots. *Nat. Hist. Mag.* **88(6)**, 60–4.

Jarman, P.J. 1974. The social organisation of antelope in relation to their ecology. *Behaviour* **48**, 215–67.

Jarvis, M.J.F. 1974. The ecological significance of clutch size in the South African gannet (*Sula capensis*, Lichenstein). *J. Anim. Ecol.* **43**, 1–17.

Jenni, D.A. 1974. Evolution of polyandry in birds. *Amer. Zool.* **14**, 129–44.

Jouventin, P. & Cornet, A. 1980. La vie sociale des phoques. *La Recherche* **105**, 1058–67.

Kacelnik, A. 1979a. The foraging efficiency of great tits (*Parus major*) in relation to light intensity. *Anim. Behav.* **27**, 237–41.

Kacelnik, A. 1979b. Studies of foraging behaviour and time-budgeting in great tits (*Parus major*). D. Phil. thesis, Oxford.

Kenward, R.E. 1978. Hawks and doves: factors affecting success and selection in goshawk attacks on wood-pigeons. *J. Anim. Ecol.* **47**, 449–60.

Kettlewell, H.B.D. 1973. *The Evolution of Melanism*. Clarendon Press, Oxford.

Kettlewell, H.B.D. & Conn, D.L. 1977. Further background choice experiments on cryptic Lepidoptera. *J. Zool. Lond.* **181**, 371–76.

Kluyver, H.N. 1971. Regulation of numbers in populations of great tit, *Parus m. major*. *Proc. Adv. Study Inst. Dynamics Numbers Popul.* (Oosterbeek), 507–23.

Knowlton, N. 1974. A note on the evolution of gamete dimorphism. *J. theor. Biol.* **46**, 283–5.

Knowlton, N. 1979. Reproductive synchrony, parental investment and the evolutionary dynamics of sexual selection. *Anim. Behav.* **27**, 1022–33.

Kodric-Brown, A. 1977. Reproductive success and the evolution of breeding territories in pupfish (*Cyprinodon*). *Evolution* **31**, 750–66.

Kramer, D.L. & Nowell, W. 1980. Central place foraging in the eastern chipmunk, *Tamias striatus*. *Anim. Behav.* **28**, 772–8.

Krebs, J.R. 1971. Territory and breeding density in the great tit, *Parus major* L. *Ecology* **52**, 2–22.

Krebs, J.R. 1978. Optimal foraging: decision rules for predators. In J.R. Krebs & N.B. Davies (eds). *Behavioural Ecology: an Evolutionary Approach*. pp. 23–63. Blackwell Scientific Publications, Oxford.

Krebs, J.R., Ashcroft, R. & Webber, M. 1978. Song repertoires and territory defence in the great tit, *Parus major*. *Nature Lond.* **271**, 539–42.

Krebs, J.R., Erichsen, J.T., Webber, M.I., Charnov, E. L. 1977. Optimal prey selection in the great tit, *Parus major. Anim. Behav.* **25**, 30–8.

Kroodsma, D.E. 1976. Reproductive development in a female song-bird: differential stimulation by quality of male song. *Science, N.Y.* **192**, 574–5.

Kroodsma, D.E. 1977. Correlates of song organization among North American wrens. *Amer. Natur.* **111**, 995–1008.

Kroodsma, D.E. 1979. Vocal duelling among male marsh wrens: evidence for ritualised expressions of dominance/subordinance. *Auk.* **96**, 506–15.

Kruuk, H. 1964. Predators and anti-predator behaviour of the black headed gull, *Larus ridibundus. Behaviour Suppl.* **11**, 1–129.

Kruuk, H. 1972. *The Spotted Hyena.* University of Chicago Press, Chicago.

Kummer, H., Götz, W. & Angst, W. 1974. Triadic differentiation: an inhibitory process protecting pair bonds in baboons. *Behaviour* **49**, 62–87.

Lack, D. 1966. *Population Studies of Birds.* Clarendon Press, Oxford.

Lack, D. 1968. *Ecological Adaptations for Breeding in Birds.* Methuen, London.

Le Boeuf, B.J. 1972. Sexual behaviour in the northern elephant seal, *Mirounga angustirostris. Behaviour* **41**, 1–26.

Le Boeuf, B.J. 1974. Male–male competition and reproductive success in elephant seals. *Amer. Zool.* **14**, 163–76.

Lewis, D. 1979. Sexual incompatibility in plants. Arnold, London.

Lewontin, R.C. 1979. Fitness survival and optimality. In D.J. Horn, R.D. Mitchell & G.R. Stairs (eds). *Analysis of Ecological Systems,* pp. 3–21. Ohio State University Press, Columbus.

Ligon, J.D. 1978. Reproductive interdependence of Piñon jays and Piñon pines. *Ecol. Monogr.* **48**, 111–26.

Lill, A. 1974. Sexual behaviour of the lek-forming white-bearded manakin. (*Manacus manacus trinitatis*). *Z. Tierpsychol.* **36**, 1–36.

Lloyd, J.E. 1979. Mating behaviour and natural selection. *Florida Entomologist* **62**, 17–23.

Lloyd, M. & Dybas, H.S. 1966. The periodical cicada problem: II Evolution. *Evolution* **20**, 466–505.

Lorenz, K. 1966. *On Aggression.* Methuen, London.

McCann, T.S. 1981. Aggression and sexual activity of male southern elephant seals. *J. Zool. Lond.* (in press).

McClintock, M.K. 1971. Menstrual synchrony and suppression. *Nature* **229**, 244–5.

McFarland, D.J. 1976. Form and function in the temporal organisation of behaviour. In P.P.G. Bateson & R.A. Hinde (eds), *Growing Points in Ethology.* pp. 55–94. Cambridge University Press, Cambridge.

McFarland, D.J. 1977. Decision making in animals. *Nature, Lond.* **269**, 15–21.

Mackinnon, J. 1974. The behaviour and ecology of wild orang-utans, *Pongo pygmaeus. Anim. Behav.* **22**, 3–74.

McPhail, J.D. 1969. Predation and the evolution of a stickleback, *Gasterosteus. J. Fish. Res. Bd. Canada* **26**, 3183–208.

Major, P.F. 1978. Predator–prey interactions in two schooling fishes, *Caranx ignobilis* and *Stolephorus purpureus. Anim. Behav.* **26**, 760–77.

Manning, A. 1961. Effects of artificial selection for mating speed in *Drosophila melanogaster. Anim. Behav.* **9**, 82–92.

Marler, P. 1955. Characteristics of some animal calls. *Nature, Lond.* **176**, 6–8.

Marten, K. & Marler, P. 1977a. Sound transmission and its significance for animal vocalization. I. Temperate habitats. *Behav. Ecol. Sociobiol.* **2**, 271–90.

Marten, K., Quine, D. & Marler, P. 1977b. Sound transmission and its significance for animal vocalization. II. Tropical forest habitats. *Behav. Ecol. Sociobiol.* **2**, 291–302.

May, R.M. 1979. When to be incestuous. *Nature, Lond.* **279**, 192–4.

Maynard Smith, J. 1964. Group selection and kin selection. *Nature, Lond.* **201**, 1145–7.

Maynard Smith, J. 1974. The theory of games and the evolution of animal conflicts. *J. theor. Biol.* **47**, 209–21.

Maynard Smith, J. 1976a. Group selection. *Q. Rev. Biol.* **51**, 277–83.

Maynard Smith, J. 1976b. Evolution and the theory of games. *Amer. Sci.* **64**, 41–5.

Maynard Smith, J. 1977a. Parental investment—a prospective analysis. *Anim. Behav.* **25**, 1–9.

Maynard Smith J. 1977b. The limitations of evolutionary theory. In R. Duncan & M. Weston-Smith (eds). *The Encyclopedia of Ignorance: Life Sciences and Earth Sciences*, pp. 235–42. Pergamon Press, Oxford.

Maynard Smith, J. 1978a. Optimization theory in evolution. *Ann. Rev. Ecol. Syst.* **9**, 31–56.

Maynard Smith, J. 1978b. *The Evolution of Sex*. Cambridge University Press, Cambridge.

Maynard Smith, J. 1978c. The ecology of sex. in J.R. Krebs & N.B Davies (eds). *Behavioural Ecology: an Evolutionary Approach*. pp. 159–79. Blackwell Scientific Publications, Oxford.

Maynard Smith, J. 1979. Game theory and the evolution of behaviour. *Proc. R. Soc. Lond. B* **205**, 475–88.

Maynard Smith, J. 1980. A new theory of sexual investment. *Behav. Ecol. Sociobiol.* **7**, 247–51.

Maynard Smith, J. 1981. The evolution of social behaviour: a classification of models. In B.C.R. Bertram, *et al.* (eds). *Current Problems in Sociobiology.* Cambridge University Press, Cambridge.

Maynard Smith, J. & Parker, G.A. 1976. The logic of asymmetric contests. *Anim. Behav.* **24**, 159–75.

Maynard Smith, J. & Price, G.R. 1973. The logic of animal conflict. *Nature, Lond.* **246**, 15–8.

Maxson, S.J. & Oring, L.W. 1980. Breeding season time and energy budgets of the polyandrous spotted sandpiper. *Behaviour,* **74**, 200–63.

Metcalf, R.A. 1980. Sex ratios, parent–offspring conflict, and local competition for mates in the social wasps *Polistes metricus* and *Polistes variatus*. *Amer. Natur.* **116**, 642–54.

Metcalf, R.A. & Finer, G. (1981). Female biased parental investment in the social wasp *Polistes apachus*. (in press).

Metcalf, R.A. & Whitt, G.S. 1977. Relative inclusive fitness in the social wasp *Polistes metricus*. *Behav. Ecol. Sociobiol.* **2**, 353–60.

Michener, C.D. 1974. *The Social Behavior of the Bees*. Belknap Press, Harvard.

Milinski, M. 1979. An evolutionarily stable feeding strategy in sticklebacks. *Z. Tierpsychol.* **51**, 36–40.

Milinski, M. & Heller, R. 1978. Influence of a predator on the optimal foraging behaviour of sticklebacks (*Gasterosteus aculeatus*) *Nature, Lond.* **275**, 642–4.

Modell, W. 1969. Horns and antlers. *Sci. Amer.* **220(4)**, 114–122.

Moehlman, P.D. 1979. Jackal helpers and pup survival. *Nature, Lond.* **277**, 382–3.

Moodie, G.E.E. 1972. Predation, natural selection and adaptation in an unusual three spined stickleback. *Heredity* **28**, 155–67.

Morris, D. 1957. 'Typical intensity' and its relation to the problem of ritualisation. *Behaviour* **11**, 1–12.

Morse, D.H. 1970. Ecological aspects of some mixed species foraging flocks of birds. *Ecol. Monogr.* **40**, 119–68.

Morton, E.S. 1975. Ecological sources of selection on avian sounds. *Amer. Natur.* **109**, 17–34.

Morton, E.S. 1977. On the occurrence and significance of motivation – structural rules in some bird and mammal sounds. *Amer. Natur.* **111**, 855–69.

Myers, J.P., Connors, P.G. & Pitelka, F.A. 1979. Territory size in wintering sanderlings: the effects of prey abundance and intruder density. *Auk* **96**, 551–61.

Myers, J.P., Connors, P.G. & Pitelka, F.A. 1981. Optimal territory size and the sanderling: comprises in a variable environment. In A.C. Kamil & T.D. Sargent (eds). *Foraging Behaviour: Ecological, Ethological and Psychological Approaches*. Garland STPM Press, New York.

Neill, S.R. St.J. &Cullen J.M. 1974. Experiments on whether schooling by their prey affects the hunting behaviour of cephalopods and fish predators. *J. Zool. Lond.* **172**, 549–69.

Nelson, J.B. 1964. Factors influencing clutch-size and chick growth in the North Atlantic gannet. *Sula bassana. Ibis* **106**, 63–77.

Nisbet, I.C.T. 1977. Courtship feeding and clutch size in common terns *Sterna hirundo*. In B. Stonehouse & C.M. Perrins (eds). *Evolutionary Ecology*, pp. 101–9. London, Macmillan.

Nottebohm, F. 1975. Continental patterns of song variability in *Zonotrichia capensis*: some possible ecological correlates *Amer. Natur.* **109**, 605–24.

O'Donald, P. 1980. *Genetic Models of Sexual Selection*. Cambridge University Press, Cambridge.

Orgel, L.E. & Crick, F.H.C. 1980. Selfish DNA: the ultimate parasite. *Nature* **284**, 604–7.

Orians, G.H. 1969. On the evolution of mating systems in birds and mammals. *Amer. Natur.* **103**, 589–603.

Orians, G.H. 1980. *Some Adaptations of Marsh-nesting Blackbirds*. Princeton University Press, Princeton.

Orians, G.H. & Pearson, N.E. 1979. On the theory of central place foraging. In D.J. Horn, R.D. Mitchell & G.R. Stairs (eds). *Analysis of Ecological Systems*, pp. 155–77. Ohio State University Press, Ohio.

Oring, L.W. 1981. Avian mating systems. In D.S. Farner & J.R. King (eds). *Avian Biology*, Vol 6. Academic Press, London (in press).

Oster, G.F. & Wilson, E.O. 1978. *Caste and Ecology In The Social Insects*. Princeton University Press, Princeton.

Packer, C. 1977 Reciprocal altruism in *Papio anubis. Nature, Lond.* **265**, 441–3.

Packer, C. 1979. Inter-troop transfer and inbreeding avoidance in *Papio anubis. Anim. Behav.* **27**, 1–36.

Page, G. & Whiteacre, D.F. 1975. Raptor predation on wintering shorebirds. *Condor* **77**, 73–83.

Parker, G.A. 1974. Assessment strategy and the evolution of animal conflicts. *J. theor. Biol.* **47**, 223–43.

Parker, G.A. 1978. Searching for mates. In J.R. Krebs & N.B. Davies (eds) *Behavioural Ecology: an Evolutionary Approach*, pp. 214–44. Blackwell Scientific Publications, Oxford.

Parker, G.A. 1979. Sexual selection and sexual conflict. In M.S. Blum & N.A. Blum (eds). *Sexual Selection and Reproductive Competition in Insects*, pp. 123–66. Academic Press, New York.

Parker, G.A., Baker, R.R. & Smith, V.G.F. 1972. The origin and evolution of gamete dimorphism and the male–female phenomenon. *J. theor. Biol.* **36**, 529–53.

Parker, G.A. & Knowlton, N. 1980. The evolution of territory size. Some ESS models. *J. theor. Biol.* **84**, 445–76.

Parker, G.A. & Thompson, E.A. 1980. Dungfly struggles: a test of the war of attrition. *Behav. Ecol. Sociobiol.* **7**, 37–44.

Partridge, L. 1980. Mate choice increases a component of offspring fitness in fruit flies. *Nature, Lond.* **283**, 290–1.

Partridge, B.L. & Pitcher, T.J. 1979. Evidence against a hydrodynamic function for fish schools. *Nature* **279**, 418–9.

Paul, R.C. & Walker, T.J. 1979. Aboreal singing in a burrowing cricket, *Anurogryllus arboreus. J. Comp Physiol.* A. **132**, 217–23.

Payne, R.B. 1977. The ecology of brood parasitism in birds. *Ann. Rev. Ecol. Syst.* **8**, 1–28.

Perril, S.A., Gerhardt, H.C. & Daniel, R. 1978. Sexual parasitism in the green tree frog, *Hyla cinerea. Science, N.Y.* **200**, 1179–80.

Perrins, C.M. 1965. Population fluctuations and clutch size in the great tit, *Parus major*. L. *J. Anim. Ecol.* **34**, 601–47.

Perrins, C.M. 1970. The timing of birds' breeding seasons. *Ibis* **112**, 242–53.

Perrins, C.M. 1979. *British Tits*. New Naturalist Series. Collins, London.

Pianka, E.R. 1974. *Evolutionary Ecology*. Harper & Row, New York.

Pitelka, F.A., Holmes, R.T. & MacLean, S.F. 1974. Ecology and evolution of social organisation in arctic sandpipers. *Amer. Zool.* **14**, 185–204.

Pleszczynska, W.K. 1978. Microgeographic prediction of polygyny in the lark bunting. *Science, N.Y.* **201**, 935–7.

Price, M.V. & Waser, N.M. 1979. Pollen dispersal and optimal outcrossing in *Delphinium nelsoni*. *Nature, Lond.* **277**, 294–7.

Prins, H.H. Th., Ydenberg, R.C. & Drent, R.H. 1980. The interaction of brent geese *Branta bernicla* and sea plantain *Plantago maritima* during spring staging: field observations and experiments. *Acta Botanica Nederlandica* (in press).

Pulliam, H.R. 1976. The principle of optimal behavior and the theory of communities. In P.H. Klopfer & P.P.G. Bateson (eds). *Perspectives in Ethology*, pp. 311–32. Plenum Press, New York.

Pulliam, H.R. & Brand, M.R. 1975. The production and utilization of seeds in plains grassland of southeastern Arizona. *Ecology* **56**, 1158–66.

Pulliam, H.R. & Dunford, C. 1980. *Programmed to learn. An essay on the evolution of culture.* Columbia University Press, N.Y.

Pyke, G.H. 1978. Optimal foraging in bumblebees and coevolution with their plants. *Oecologia* **36**, 281–93.

Pyke, G.H. 1979a. The economics of territory size and time budget in the golden-winged sunbird. *Amer. Natur.* **114**, 131–45.

Pyke, G.H. 1979b. Optimal foraging in bumblebees: rule of movement between flowers within inflorescences. *Anim. Behav.* **27**, 1167–81.

Pyke, G.H., Pulliam, H.R. & Charnov, E.L. 1977. Optimal foraging: a selective review of theory and tests. *Q. Rev. Biol.* **52**, 137–54.

Ralls. K., Brugger, K. & Ballou, J. 1979. Inbreeding and juvenile mortality in small populations of ungulates. *Science, N.Y.* **206**, 1101–3.

Reyer, H.-U. 1980. Flexible helper structure as an ecological adaptation in the pied kingfisher, *Ceryle rudis rudis*. L. *Behav. Ecol. Sociobiol.* **6**, 219–27.

Ridley, M. 1978. Paternal care. *Anim. Behav.* **26**, 904–32.

Riechert, S.E. 1978. Games spiders play: behavioural variability in territorial disputes. *Behav. Ecol. Sociobiol.* **3**, 135–62.

Riechert, S.E. 1979. Games spiders play. II. Resource assessment strategies. *Behav. Ecol. Sociobiol.* **6**, 121–8.

Robertson, D.R., Sweatman, H.P.A., Fletcher, E.A. & Cleland, M.G. 1976. Schooling as a Mechanism for circumventing the territoriality of competitors. *Ecology* **57**, 1208–30.

Rohwer, S. & Rohwer, F.C. 1978. Status signalling in Harris sparrows: experimental deceptions achieved. *Anim. Behav.* **26**, 1012–22.

Rood, J.P. 1978. Dwarf mongoose helpers at the den. *Z. Tierpsychol.* **48**, 277–87.

Rosenzweig, M.L & MacArthur, R.H. 1963. Graphical representation and stability conditions of predator–prey interactions. *Amer. Natur.* **97**, 209–23.

Rothenbuhler, W.C. 1964. Behaviour genetics of nest cleaning in honey bees. IV. Responses of F1 and backcross generations to disease killed brood. *Amer. Zool.* **4**, 111–23.

Rothstein, S.I. 1979. Gene frequencies and selection for inhibitory traits, with special emphasis on the adaptiveness of territoriality. *Amer. Natur.* **113**, 317–31.

Rubenstein, D.I. 1978. On predation, competition, and the advantages of group living. In P.P.G. Bateson & P.H. Klopfer. *Perspectives in Ethology*, pp. 205–31. Plenum Press, New York.

Rubenstein, D.I. 1980. On the evolution of alternative mating strategies. In J.E.R. Staddon (ed.). *Limits to Action: The Allocation of Individual Behaviour.* Academic Press, New York.

Sargent, T.D. 1968. Cryptic moths: effects on background selection of painting the circumocular scales. *Science, N.Y.* **159**, 100–1.

Sauer, E.G.F. & Sauer, E.M. 1959. Polygamie beim Südafrikanischen Strauss. *Bonn. zool. Beitr.* **10**, 266–85.

Seghers, B.H. 1974. Schooling behaviour in the guppy *Poecilia reticulata*: an evolutionary response to predation. *Evolution* **28**, 486–9.

Selander, R.K. 1972. Sexual selection and dimorphism in birds. In B. Campbell (ed.) *Sexual Selection and the Descent of Man*, pp. 180–230. Aldine, Chicago.

Semler, D.E. 1971. Some aspects of adaptation in a polymorphism for breeding colours in the three-spine stickleback. (*Gasterosteus aculeatus* L.) *J. Zool. Lond.* **165**, 291–302.

Shalter, M.D. 1978. Localisation of passerine seet and mobbing calls by goshawks and pygmy owls. *Z. Tierpsychol.* **46**, 260–7.

Shepher, J. 1971. Mate selection among second generation kibbutz adolescents and adults: incest avoidance and negative imprinting. *Arch. sex. Behav.* **1**, 293–307.

Sherman, P.W. 1977. Nepotism and the evolution of alarm calls. *Science, N.Y.* **197**, 1246–53.

Sherman, P.W. 1981. Reproductive competition and infanticide in Belding's ground squirrels and other animals. In R.D. Alexander & D.W. Tinkle (eds). *Natural Selection and Social Behaviour: Recent Research and New Theory.* Chiron Press, New York.

Sherman, P.W. 1980. The limits of ground squirrel nepotism. In G.W. Barlow & J. Silverberg (eds). *Sociobiology: Beyond Nature/Nurture*, pp. 505–44. Westview Press, Boulder, Colorado.

Silverman, H.B. & Dunbar, M.J. 1980. Aggressive tusk use by the narwhal, *Monodon monoceros*, L. *Nature, Lond.* **284**, 57–8.

Silverin, B. 1980. Effects of long-acting testosterone treatment on free-living pied flycatchers *Fidecula hypoleuca. Anim. Behav.* **28**, 906–12.

Silverstone, T. (ed.) 1976. *Appetite and food intake.* Dahlem Workshop life Sciences Res. Report vol 2. Verlag Chemie, Weinheim.

Simon, C. 1979. Debut of the seventeen-year-old cicada. *Nat. Hist.* **88**(5), 38–45.

Simpson, M.J.A. 1968. The display of Siamese fighting fish, *Betta splendens. Anim. Behav. Monogr.* **1**, 1–73.

Sinclair, A.R.E. 1977. *The African Buffalo.* University of Chicago Press, Chicago.

Slatkin, M. & Maynard Smith, J. 1979. Models of coevolution. *Q. Rev. Biol.* **54**, 233–63.

Smith, N.G. 1968. The advantage of being parasitized. *Nature, Lond.* **219**, 690–4.

Smith, R.L. 1979. Repeated copulation and sperm precedence: Paternity assurance for a male brooding water bug. *Science, N.Y.* **205**, 1029–31.

Snow, D.W. 1971. Evolutionary aspects of fruit-eating by birds. *Ibis* **113**, 194–202.

Stacey, P.B. 1979. Habitat saturation and communal breeding in the acorn woodpecker. *Anim. Behav.* **27**, 1153–66.

Staddon, J.E.R. 1980. Optimality analyses of operant behaviour and their relation to optimal foraging. In J.E.R. Staddon (ed.). *Limits to action. The allocation of individual behavior.* Academic Press, London.

Stern, D. 1977. *The First Relationship: Infant and Mother.* Fontana Open Books, London.

Stokes, A.W. 1962. Agonistic behaviour among blue tits at a winter feeding station. *Behaviour* **19**, 118–38.

Struhsaker, T. 1975. *The Red Colobus Monkey.* Chicago University Press, Chicago.

Taborsky, M. & Limberger, D. 1981. Helpers in fish. *Behav. Ecol. Sociobiol.* (in press).

Thornhill, R. 1976. Sexual selection and nuptial feeding behaviour in *Bittacus apicalis* (Insecta: Mecoptera). *Amer. Natur.* **110**, 529–48.

Thornhill, R. 1979. Adaptive female-mimicking behavior in a scorpionfly. *Science, N.Y.* **205**, 412–4.

Thornhill, R. 1980. Rape in *Panorpa* scorpion flies and a general rape hypothesis. *Anim. Behav.* **28**, 52–9.

Tinbergen, N. 1953. *The Herring Gull's World.* New Naturalist Series, Collins, London.

Tinbergen, N. 1957. The functions of territory. *Bird Study* **4**, 14–27.

Tinbergen, N. 1963. On aims and methods of ethology. *Z. Tierpsychol.* **20**, 410–33.

Tinbergen, N., Impekoven, M. & Franck, D. 1967. An experiment on spacing out as a defence against predators. *Behaviour* **28**, 307–21.

Trivers, R.L. 1971. The evolution of reciprocal altruism. *Q. Rev. Biol.* **46**, 35–57.

Trivers, R.L. 1972. Parental investment and sexual selection. In B. Campbell (ed.). *Sexual Selection and the Descent of Man*, pp. 139–79. Aldine, Chicago.

Trivers, R.L. 1974. Parent–offspring conflict. *Amer. Zool.* **14**, 249–64.

Trivers, R.L. & Hare, H. 1976. Haplodiploidy and the evolution of social insects. *Science, N.Y.* **191**, 249–63.

Trune, D.R. & Slobodchikoff, C.N. 1976. Social effects of roosting on the metabolism of the pallid bat, *Antrozous pallidus*. J. Mammal. **57**, 656–63.

Vander Wall, S.B. & Balda, R.P. 1977. Coadaptations of the Clark's nutcracker and the pinon pine for efficient seed harvest and dispersal. *Ecol. Monogr.* **47**, 89–111.

van Rhijn, J.G. 1973. Behavioural dimorphism in male ruffs *Philomachus pugnax* (L.). *Behaviour* **47**, 153–229.

van Valen, L.M. 1980. Patch selection, benefactors and a rivitalisation of ecology. *Evolutionary Theory* **4**, 231–3.

Vehrencamp, S.L. 1977. Relative fecundity and parental effort in communally nesting anis, *Crotophaga sulcirostris*. *Science, N.Y.* **197**, 403–5.

Venhrencamp, S.L. 1978. The adaptive significance of communal nesting in groove-billed anis, *Crotophaga sulcirostris*. *Behav. Ecol. Sociobiol.* **4**, 1–33.

Vehrencamp, S.L. 1980. The roles of individual, kin and group selection in the evolution of sociality, pp. 351–94. In P. Marler & J.G. Vandenbergh (eds). *Handbook of Behavioral Neurobiology vol 3. Social behavior and communication*. Plenum Press, New York.

Verner, J. 1964. Evolution of polygamy in the long-billed marsh wren. *Evolution* **18**, 252–61.

Verner, J. 1977. On the adaptive significance of territoriality. *Amer. Natur.* **111**, 769–75.

Verner, J. & Willson, M.F. 1966. The influence of habitats on mating systems of North American passerine birds. *Ecology* **47**, 143–7.

Vogel, S., Ellington, C.P. & Kilgore, D.L. 1973. Wind-induced ventilation of the burrows of the prairie dog *Cynomys ludovicianus J. Comp. Physiol.* **85**, 1–14.

Waage, J.K. 1979. Dual function of the damselfly penis: sperm removal and transfer. *Science, N.Y.* **203**, 916–8.

Waldman, B. & Adler, K. 1979. Toad tadpoles associate preferentially with siblings. *Nature, Lond.* **282**, 611–3.

Ward, P. & Zahavi, A. 1973. The importance of certain assemblages of birds as 'information-centres' for food finding. *Ibis* **115**, 517–34.

Warner, R.R. 1975. The adaptive significance of sequential hermaphroditism in animals. *Amer. Natur.* **109**, 61–82.

Warner, R.R. 1978. The evolution of hermaphroditism and unisexuality in aquatic and terrestrial vertebrates. In E.S. Reese & F.J. Lighter (eds). *Contrasts In Behaviour*, pp. 77–101. Wiley, N.Y.

Warner, R.R. & Hoffman, S.G. 1980. Local population size as a determinant of mating system and sexual composition in two tropical marine fishes (*Thalassoma* spp.) *Evolution* **34**, 508–18.

Warner, R.R., Robertson, D.R. & Leigh, E.G. 1975. Sex change and sexual selection. *Science N.Y.* **190**, 633–8.

Waser, N.M. 1978. Competition for hummingbird pollination and sequential flowering in two Colorado wildflowers. *Ecology* **59**, 934–44.

Waser, P.M. & Waser, M.S. 1977. Experimental studies of primate vocalization: specializations for long distance propagation. *Z. Tierpsychol.* **43**, 239–63.

Watson, A. 1967. Territory and population regulation in the red grouse. *Nature, Lond.* **215**, 1274–5.

Watson, A. 1970. Territorial and reproductive behaviour of red grouse. *J. Reprod. Fert. Suppl.* **11**, 3–14.

Watts, C.R. & Stokes, A.W. 1971. The social order of turkeys. *Sci. Amer.* **224 (6)**, 112–8.

Weihs, D. 1973. Hydromechanics of fish schooling. *Nature* **241**, 290–1.

Wells, K.D. 1977. The social behaviour of anuran amphibians. *Anim. Behav.* **25**, 666–93.

Werren, J.H. 1980. Sex ratio adaptations to local mate competition in a parasitic wasp. *Science, N.Y.* **208**, 1157–9.

West Eberhard, M.J. 1975. The evolution of social behaviour by kin selection. *Q. Rev. Biol.* **50**, 1–33.

West Eberhard, M.J. 1978a. Temporary queens in *Metapolybia* wasps: non-reproductive helpers without altruism? *Science, N.Y.* **200**, 441–3.

West Eberhard, M.J. 1978b. Polygyny and the evolution of social behavior in wasps. *J. Kans. Ent. Soc.* **51**, 832–56.

Whitham, T.G. 1977. Coevolution of foraging in *Bombus*—nectar dispensing in *Chilopsis*: a last dreg theory. *Science, N.Y.* **197**, 593–6.

Whitham, T.G. 1978. Habitat selection by *Pemphigus* aphids in response to resource limitation and competition. *Ecology* **59**, 1164–76.

Whitham, T.G. 1979. Territorial behaviour of *Pemphigus* gall aphids. *Nature, Lond.* **279**, 324–5.

Whitham, T.G. 1980. The theory of habitat selection examined and extended using *Pemphigus* aphids. *Amer. Natur.* **115**, 449–66.

Whitney, C.L. & Krebs, J.R. 1975a. Mate selection in Pacific tree frogs. *Nature, Lond.* **255**, 325–6.

Whitney, C.L. & Krebs, J.R. 1975b. Spacing and calling in Pacific tree frogs, *Hyla regilla. Can. J. Zool.* **53**, 1519–27.

Wickler, W. 1968. *Mimicry in Plants and Animals.* World University Library, London.

Wiley, R.H. 1973. Territoriality and non-random mating in the sage grouse, *Centrocercus urophasianus. Anim. Behav. Monogr.* **6**, 87–169.

Wiley, R.H. & Richards, D.G. 1978. Physical constraints on acoustic communication in the atmosphere: implications for the evolution of animal vocalizations. *Behav. Ecol. Sociobiol.* **3**, 69–94.

Wilkinson, P.F. & Shank, C.C. 1977. Rutting-fight mortality among musk oxen on Banks Island, Northwest Territories, Canada. *Anim. Behav.* **24**, 756–8.

Williams, G.C. 1966. *Adaptation and Natural Selection.* Princeton University Press, Princeton.

Williams, G.C. 1979. The question of adaptive sex ratio in outcrossed vertebrates. *Proc. R. Soc. Lond. B.* **205**, 567–80.

Wilson, D.S. 1980. *The Natural Selection of Populations and Communities.* Benjamin/Cummings, Menlo Park, California.

Wilson, E.O. 1971. *The Insect Societies.* Belknap Press, Harvard.

Wilson, E.O. 1975. *Sociobiology: The New Synthesis.* Belknap Press, Harvard.

Wilson, E.O. 1980. Caste and division of labor in leaf-cutter ants (Hymenoptera Formicidae: *Atta*) II. The ergonomic optimization of leaf cutting. *Behav. Ecol. Sociobiol.* **7**, 157–65.

Wittenberger, J.F. 1980. Group size and polygamy in social mammals. *Amer. Natur.* **115**, 197–222.

Woolfenden, G.E. 1975. Florida scrubjay helpers at the nest. *Auk* **92**, 1–15.

Woolfenden, G.E. & Fitzpatrick, J.W. 1978. The inheritance of territory in group breeding birds. *BioScience* **28**, 104–8.

Wrangham, R.W. 1980. An ecological model of female-bonded primate groups. *Behaviour* **75**, 262–300.

Wynne-Edwards, V.C. 1962. *Animal Dispersion in Relation to Social Behaviour.* Oliver & Boyd, Edinburgh.

Yasukawa, K. 1979. A fair advantage in animal confrontations. *New Sci.* **84**, 366–8.

Yom-Tov, Y. 1980. Intraspecific nest parasitism in birds. *Biol. Rev.* **55**, 93–108.

Zach, R. 1979. Shell dropping: decision making and optimal foraging in Northwestern crows. *Behaviour* **68**, 106–17.

Zahavi, A. 1974. Communal nesting by the Arabian babbler: a case of individual selection. *Ibis* **116**, 84–7.

Zahavi, A. 1975. Mate selection – a selection for a handicap. *J. theor. Biol.* **53**, 205–14.

Zahavi, A. 1977. The cost of honesty (further remarks on the handicap principle). *J. theor. Biol.* **67**, 603–5.

Zahavi, A. 1979. Ritualisation and the evolution of movement signals. *Behaviour* **72**, 77–81.

Zuckerman, S. 1932. *The Social Life of Monkeys and Apes.* Paul, Trench, Trubner & Co. London.

Author index

Subject index

285